Canadian Bulletin of Fisheries and Aquatic Sciences 227

Atlantic Halibut (*Hippoglossus hippoglossus*) and Pacific Halibut (*H. stenolepis*) and their North American Fisheries

Robert J. Trumble

International Pacific Halibut Commission
P.O. Box 95009, Seattle, Washington 98145-2009, U.S.A.

John D. Neilson

Department of Fisheries and Oceans
St. Andrews Biological Station, St. Andrews, New Brunswick, Canada E0G 2X0

W. R. Bowering

Department of Fisheries and Oceans
P.O. Box 5667, St. John's, Newfoundland, Canada A1C 5X1

Donald A. McCaughran

International Pacific Halibut Commission
P.O. Box 95009, Seattle, Washington 98145-2009, U.S.A.

NATIONAL RESEARCH COUNCIL OF CANADA
Ottawa 1993

ISBN 0-660-15111-1
ISSN 0706-6503
NRCC No. 36177

Canadian Cataloguing In Publication Data

Trumble, R.J., Neilson, J.D., Bowering, W.R., and McCaughran, D.A. Atlantic halibut (*Hippoglossus hippoglossus*) and Pacific halibut (*H. stenolepis*) and their North American fisheries

(Canadian Bulletin of Fisheries and Aquatic Sciences; No. 227)
Includes an abstract in French. Includes bibliographical references.
ISBN 0-660-15111-1
DSS Cat. No. Fs94-227E

1. Halibut fisheries -- North America. 2. Atlantic halibut. 3. Pacific halibut.
I. National Research Council Canada. II. Trumble, Robert J. III. Series.

SH351.H2A84 1993 333.95'6 C93-099556-2

This publication is available from:

Subscription Office, Research Journals, National Research Council of Canada, Building M-55, Montreal Road, Ottawa, Ontario, Canada K1A 0R6

Enquiries: Tel.: 613-993-9084 FAX: 613-952-7656

Remittances should be made payable to the Receiver General for Canada, credit National Research Council of Canada.

Research Journals, NRC
Editor-in-chief Bruce P. Dancik

Publishing Department, NRC
Director Aldyth Holmes
Editorial and Publishing Services Gerald J. Neville/Mike Boroczki

---

Printed on recycled paper.

The views expressed are those of the authors and are not necessarily those of the National Research Council of Canada.

Cover drawing courtesy of Joan Forsberg, IPHC

Correct citation for this publication:

Trumble, R.J., Neilson, J.D., Bowering, W.R., and McCaughran, D.A. 1993. Atlantic halibut (*Hippoglossus hippoglossus*) and Pacific halibut (*H. stenolepis*) and their North American fisheries. Can. Bull. Fish. Aquat. Sci. 227: 84 pp.

Contents

Abstract

Trumble, R. J., J. D. Neilson, W. R. Bowering, and D. A. McCaughran. 1993. Atlantic halibut (*Hippoglossus hippoglossus*) and Pacific halibut (*H. stenolepis*) and their North American fisheries. Can. Bull. Fish. Aquat. Sci. 227: 84 pp.

Atlantic halibut (*Hippoglossus hippoglossus*) and Pacific halibut (*H. stenolepis*) off the Atlantic and Pacific coasts of North America share many biological characteristics. The conventional separation of the halibut into separate species is not uniformly accepted. Reproduction and early life history are nearly identical, although the extensive drift of Pacific halibut eggs and larvae and countermigration as juveniles have not been demonstrated for the Atlantic halibut. Adult halibut undergo migration from winter spawning grounds along the upper continental slope to summer feeding grounds on the continental shelf. The two halibuts show the same pattern of age and growth. They exhibit different trophic selectivity, as small Atlantic halibut consume a higher proportion of invertebrates as small fish and large Atlantic halibut consume a higher proportion of fish, compared to Pacific halibut. Halibut management in the Atlantic and Pacific varies dramatically. Atlantic halibut in Canada were managed with other groundfish until 1988 and are not managed in the United States. Pacific halibut have been managed by an international commission under treaty between the United States and Canada since 1923. Atlantic halibut have produced from 1000 to 5000 t annually, mainly off Canada, while Pacific halibut have produced 12 000 to 45 000 t, mainly off Alaska. Biological and fishery data are collected more extensively for the Pacific halibut, for which detailed stock assessment is conducted.

Résumé

Le flétan de l'Atlantique (*Hippoglossus hippoglossus*) et le flétan du Pacifique (*H. stenolepis*) qui vivent au large des côtes de l'Atlantique et du Pacifique de l'Amérique du Nord ont de nombreuses caractéristiques biologiques en commun. La séparation traditionnelle du flétan en espèces distinctes n'est pas uniformément acceptée. La reproduction et les premiers stades de vie sont presque identiques, bien que la dérive sur une grande distance des oeufs et des larves de flétan du Pacifique et la contre-migration des juvéniles n'aient pas été mises en évidence dans le cas du flétan de l'Atlantique. Les flétans adultes migrant depuis les aires de frai d'hiver situées le long de la partie supérieure du talus continental pour se rendre dans les aires d'alimentation d'été situées sur le plateau continental. Les deux espèces de flétan montrent les caractéristiques semblables du point de vue de l'âge et de la croissance. Elles présentent une sélectivité trophique différent du fait que, comparativement au flétan du Pacifique, le flétan de l'Atlantique de petite taille consomme une proportion plus élevée d'invertébrés et le flétan de l'Atlantique de grande taille consomme une proportion plus élevée de poissons. La gestion du flétan dans l'Atlantique et dans le Pacifique est totalement différente. Au Canada, la gestion du flétan de l'Atlantique était intégrée à celle des autres poissons de fond jusqu'en 1988, et aux États-Unis, la gestion est inexistante. Par contre, le flétan du Pacifiques est géré depuis 1923 par une commission internationale en vertu d'un traité intervenue entre le Canada et les États-Unis. Les débarquements de flétan de l'Atlantique varient entre 1000 et 5000 t par année, les prises étant effectuées principalement au large du Canada, et les débarquements de flétan du Pacifique varient de 12 000 à 45 000 t par année, les prises étant effectuées principalement au large de lAlaska. La collecte de données biologiques et halieutiques est plus poussée dans le cas du flétan du Pacifique dont les stocks font l'objet d'une évaluation détaillée.

Introduction

Atlantic halibut (*Hippoglossus hippoglossus*) and Pacific halibut (*H. stenolepis*) are closely related taxa that share many biological similarities (Fig. 1). Both are highly valued groundfish that support significant commercial fisheries. In general, annual Pacific halibut landings are greater than those from the Atlantic fishery by about an order of magnitude. Management of halibut differs considerably on each coast, offering a unique opportunity to examine the effects of different management regimes and harvesting strategies on the population dynamics of closely related (possibly conspecific; Bell 1981) taxa.

Pacific halibut have been managed since 1923 when the International Fisheries Commission, now called the International Pacific Halibut Commission (IPHC), was formed by treaty between the U.S. and Canada to manage halibut in the waters adjacent to the two countries. The IPHC collected biological and environmental data since its inception and has a data base that includes catch records from the turn of the century, standardized effort and age composition since 1935, tagging results since 1925, and results of various special projects conducted throughout the history of the Commission. Pacific halibut is considered as a single population throughout its range but is managed by regulatory area. Strict size limits and gear restrictions are applied. Abundance and harvest have fluctuated since the beginning of the fishery but are currently at near record high levels.

In contrast to the Pacific situation, the Atlantic halibut fishery in North America was unregulated until 1988 in Canada. To some extent, the lack of active management in the past reflected the greater interest by industry in the more abundant Atlantic groundfish resources such as Atlantic cod (*Gadus morhua*) and haddock (*Melanogrammus aeglefinus*). Recently, however, the introduction of more efficient harvesting techniques, higher prices, and reduced quotas for the more traditional groundfish have led to increased pressure on the Atlantic halibut resource and initiation of management measures by Canada. Two Canadian stocks are thought to exist (Neilson et al. 1987): a comparatively small one in the Gulf of St. Lawrence and a larger stock encompassing the Scotian Shelf and the southern Grand Banks of Newfoundland. The Atlantic halibut fishery in the United States is practically extinct.

In this paper, we present a comprehensive review of the biology and exploitation of the two halibuts. The focus of the review is on halibut of North America (Figs. 2*a* and 2*b*), but data from other regions are included where appropriate. By establishing the level of similarity in the biology and fisheries of the two taxa (Table 1), we believe that a better understanding of the possible utility of management measures for either population can be gained.

Fig. 1*a*. Atlantic halibut.

Fig. 1*b*. Pacific halibut.

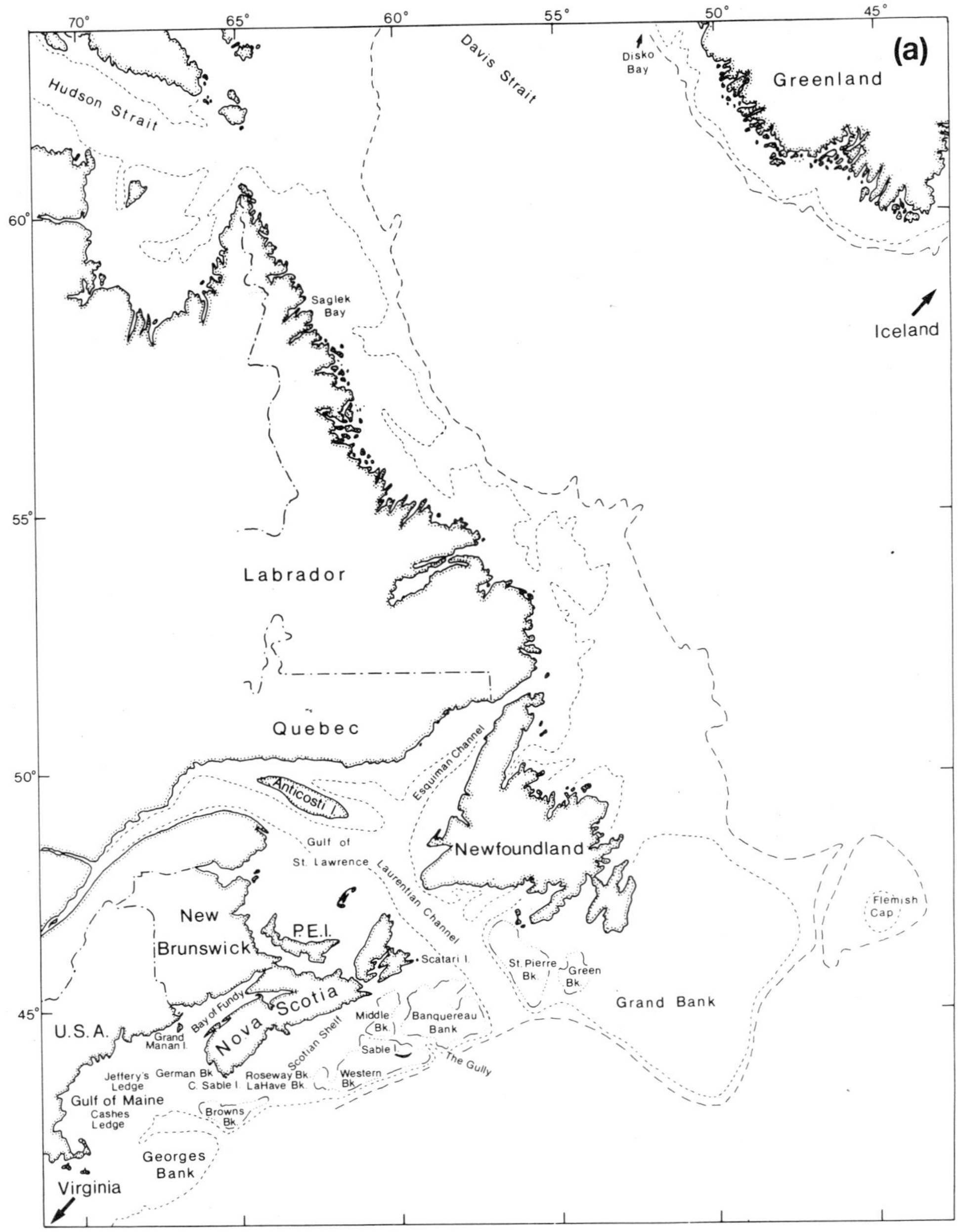

Fig. 2*a*. Geography of Atlantic Ocean areas and place names referred to in the text.

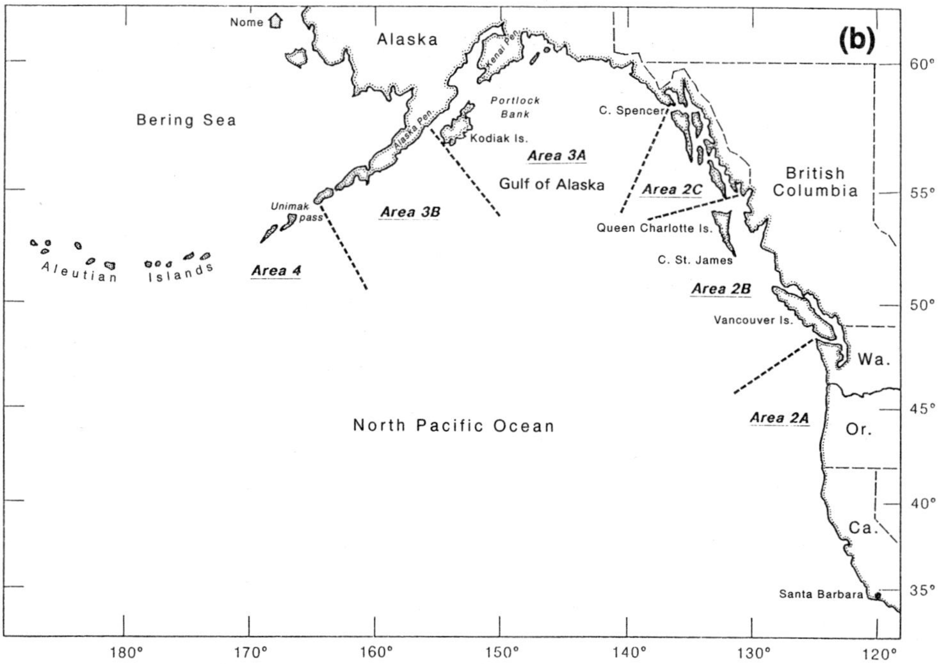

Fig. 2*b*. Geography of Pacific Ocean areas and place names referred to in text.

Table 1. Comparison of biological and fishery characteristics of Pacific and Atlantic halibut

Characteristic	Pacific	Atlantic	Remarks
		Biology	
Taxonomy	*H. stenolepis*	*H. hippoglossus*	Specific status not conclusively demonstrated
Spawning season	November–March	February–April	Atlantic spawning duration under study
Spawning area	Upper continental slope	Upper continental slope	Atlantic spawning areas not well known
Spawning temperature (°C)	3–8	3–8	
Eggs			
Diameter (mm)	2.9–3.8	2.9–3.5	
Density	Neutral	Neutral	
Age at 50% maturity (years)			
Female	12	12	
Male	8	8	
Maximum age (years)			
Female	42	30–35 *	* Sex not indicated for Atlantic (Scott and Scott 1988)
Male	27		
Size at 50% maturity (cm)			
Female	119	119	Pacific data for males may contain errors (see text)
Male	Uncertain	77	

Table 1 (*concluded*)

Characteristic	Pacific	Atlantic	Remarks
		Fishery	
Natural mortality	0.2	0.2	Atlantic value assumed from Pacific value
Range	Northern California to Bering Sea	Virginia to Disko Bay, Greenland	Distribution continues to Europe in the Atlantic and to Japan in the Pacific
Migration	To 1000's of miles	To 1000's of miles	Net migration as juveniles, spawning migration as adults
Removals (990), t			
Commercial	37 000	2400	
Sport	3700	None	
Bycatch	10 500	Unknown	
Waste	2000	Unknown	
Peak harvest (t)		4200 (1985)	"Total" includes noncommercial removals
Commercial	45 300 (1974)		
Total			
10 Year average yield (t)			
Commercial	27 300	3000	
Total	40 200	Unknown	
Stock size (1990), t			
Biomass	141 900	Unknown	
Price per pound (1990)	$1.88 US	$4.16 US*	*Canadian price for Nova Scotia only
Active vessels (1990)	4699	1886	U.S. and Canada in the Pacific, Canada only in Atlantic
		Management	
Gear restrictions	Retention by hook and line only	Trawl retention permitted	
Size limits (cm)	81.3	81.3	Commercial limit; no size limit on sport fishing except in Oregon
Catch limits (1990), t	35 400	3500	Commercial limit only
Stock assessment	Catch-at-age analysis	Catch per unit effort	
Stock trend	High but declining	Low and declining	Nearly absent in Atlantic off US
Budget	$1.5 million US	$50 000 US	

Biological Review

Taxonomic Status

The taxonomic position of the two halibuts is not conclusive, as they have been described as a single species, subspecies, and as separate species. The generally accepted taxonomic positions of Atlantic and Pacific halibut are shown in Fig. 3.

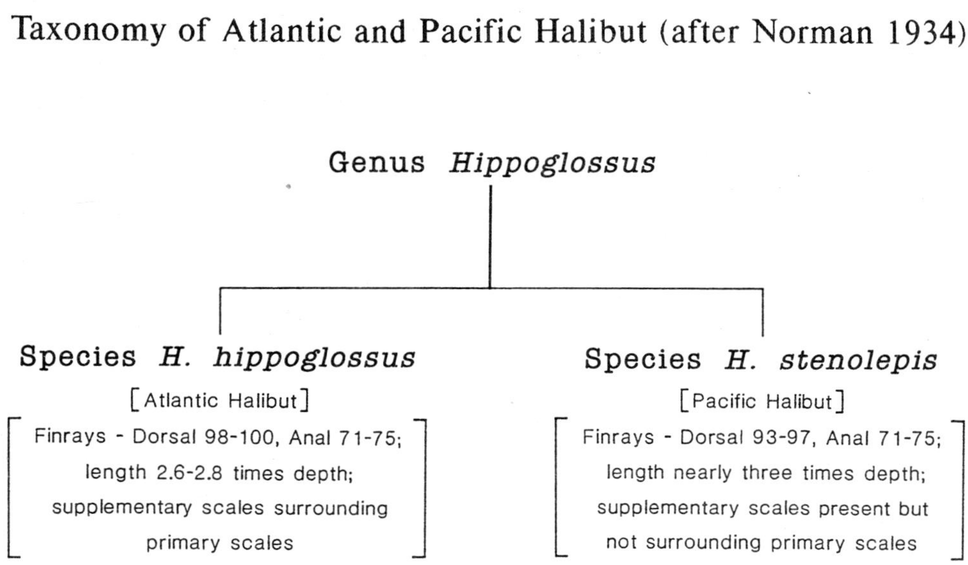

Fig. 3. Taxonomy of Atlantic and Pacific halibut (after Norman 1934).

The Atlantic halibut was first described by Carolus Linnaeus in 1758 as *Pleuronectes hippoglossus* (Norman 1934; Bell 1981). At this time, all flatfishes known to Linnaeus were placed in the single genus *Pleuronectes*. It was more than 50 years later that Pallas (1814) first used this same name for halibut in the northwest Pacific Ocean in the vicinity of the Kamchatka peninsula of the Soviet Union. Cuvier's (1817) classification of the flatfishes raised them to the rank of "family" and associated them with several other groups of fishes, based on the thoracic position of the pelvic fins and the absence of spines in the dorsal fin (Norman 1934). During this period, he introduced the genus *Hippoglossus* and renamed the halibut *Hippoglossus hippoglossus*. Fleming (1828) altered the name to *Hippoglossus vulgaris* and this name was used to refer to both Atlantic and Pacific halibuts until 1885 when Jordan re-introduced *Hippoglossus hippoglossus* for both forms (Bell 1981).

It was not until 1904 that the Pacific halibut was described as a separate species. P. J. Schmidt (cited in Bell 1981), after examining halibut from the vicinity of the Sakhalin Island in the northwest Pacific, named it *Hippoglossus stenolepis*. In 1930, Schmidt compared halibut from this area and those of the eastern Pacific Ocean and concluded they were the same species, while halibut from the Atlantic Ocean were different. His investigations were based mainly upon meristic characters and, more importantly, the morphology of the scales. Scale morphology led him to suggest the specific name *stenolepis*, with "steno" meaning narrow and "lepis" meaning scale (Bell 1981). The comparison was based on only seven specimens from the Atlantic and three from the Pacific. As Bell (1981) points out, this is an unacceptably low sample size for distinguishing species. A year after Schmidt's work, Rendahl (cited in Norman 1934) compared a single Pacific specimen from near the Kamchatka peninsula with several from the Northeast Atlantic and concluded it was intermediate. He therefore named it *Hippoglossus hippoglossus camchaticus*.

Subsequent studies by Vernidub (1936) and unpublished studies by F. H. Bell for the IPHC (summarized in Bell 1981) used meristic and morphometric data to conclude that there was little evidence for supporting the Pacific halibut as a species separate from the Atlantic halibut. Vernidub and Bell asserted that meristic counts overlapped for Atlantic and Pacific halibut and attributed any measured variations to stock, rather than to species, level differences. However, readers must use Bell's results cautiously, as his meristic analyses were not published or scrutinized in the scientific literature, and inconsistencies appear. For example, the only overlap in counts indicted in Bell's Table 2 is from an Atlantic sample with an unexplained correction "for an anomalous count." Otherwise, data presented in Bell's Tables 2 and 51 show no overlap in counts, and results are similar to those of Schmidt.

Many fish stock identification studies have used meristic and morphometric characteristics and noted the confounding effects of environment and genetics, even at the stock level. For example, Bowering (1988) concluded that a multivariate morphometric analysis of Greenland halibut[1] (*Reinhardtius hippoglossoides*) would have erroneously distinguished eight different stocks in the northwest Atlantic study area where it was already known that only one existed (Fairbairn 1981; Khan et al. 1982; Misra and Bowering 1984). The contradiction was probably due to north-south clinal variation in morphometric measurements of the specimens examined which were taken sufficiently far apart to be significantly different from each other. Lear and Misra (1978) similarly showed north-south clinal variation in scale characteristics of Atlantic salmon (*Salmo salar*) from 18 river systems in eastern North America. Given the distribution of halibut throughout the Pacific and Atlantic Oceans, it would not be unreasonable to expect such variations in meristic and morphometric characteristics. Given the conclusions of the workers mentioned above, it would appear that morphometric data are not adequate to determine if different species occur in the Atlantic and Pacific halibut.

A more recent study by Grant et al. (1984) used electrophoretic techniques to examine possible genetic differences between Atlantic and Pacific halibut. According to their measurements of genetic distance between samples of Pacific and Atlantic halibut, the ancestors of Atlantic halibut moved into the North Atlantic from the Pacific between 4.5 and 1.7 million years ago. This would include the Pliocene period of the formation of the Bering Strait. They further concluded that, by the

---

[1] The American Fisheries Society accepts Greenland halibut as the official common name but noted that the official market name in the U.S. and Canada is Greenland turbot.

criteria of Shaklee et al. (1982), their "estimate of biochemical genetic distance between Pacific and Atlantic halibut is within the range of genetic distances between species and confirms the practice of treating these taxa as species."

Grant et al. (1984) electophoretically analyzed 35 protein-coding loci of juvenile Pacific and Atlantic halibut and reviewed the limited number of prior publications that addressed halibut genetics. Of the loci tested, 17 (49%) for Pacific halibut and 6 (17%) for Atlantic were polymorphic, but only 5 and 3 loci for Pacific and Atlantic halibut, respectively, had variant allele frequencies large enough to test for population structure (greater than 0.05). No significant differences were found to indicate population differences between Gulf of Alaska and Bering Sea halibut, but a significant difference was found at one locus between a Japanese sample and the pooled Gulf of Alaska and Bering Sea samples.

Tsuyuki et al. (1969) also showed genetic homogeneity for juvenile halibut in the Gulf of Alaska and Bering Sea, with the exception of a distinct population in southeast Alaska. Grant et al. (1984) discounted Tsuyuki et al.'s (1969) conclusion of a separate population in southeast Alaska, arguing that statistical tests were not correctly applied.

Grant et al. (1984) recognized the weakness of applying genetic analysis only to juvenile halibut, but concluded that results of their studies and those of Tsuyuki et al. (1969) were consistent with a larval drift and juvenile migration model (see sections on Early Life History and Juvenile Migration), in which not all juvenile halibut return to their natal area. However, the potential exists for finding separate populations by limiting analysis to adults, if the highly mixed juveniles segregate as adults to natal areas during migration.

The Grant et al. (1984) paper, while substantially advancing our understanding of the taxonomic status of Atlantic and Pacific halibut, had a number of limitations. For example, they do not cite the sample sizes. Furthermore, their estimate of the lower confidence bound of the biochemical genetic distance (0.089) was fairly close to that expected from intraspecific populations (0.002-0.065), according to Shaklee et al. (1982). Perhaps most importantly, we question whether the Icelandic sample employed by Grant et al. (1984) is representative of the Atlantic population as a whole. Fevolden and Haug (1988) found no evidence for genetically distinct races of Atlantic halibut in the North Atlantic. However, no material from populations of Atlantic halibut west of Greenland was included in their comparison. Fevolden and Haug's conclusion of the absence of separate races is inconsistent with McCracken (1958), who noted that on the basis of differences in movements, size and age composition of the commercial landings, and growth rate, the Gulf of St. Lawrence halibut population was probably discrete from the Scotian Shelf population. Indeed, current management of Canadian Atlantic halibut reflects this apparent difference (Neilson et al. 1987). Hence, the possibility that separate races occur in Atlantic halibut cannot yet be dismissed, and the validity of the assumption of Grant et al. (1984) that the Icelandic sample is representative of the North Atlantic genome seems open to question.

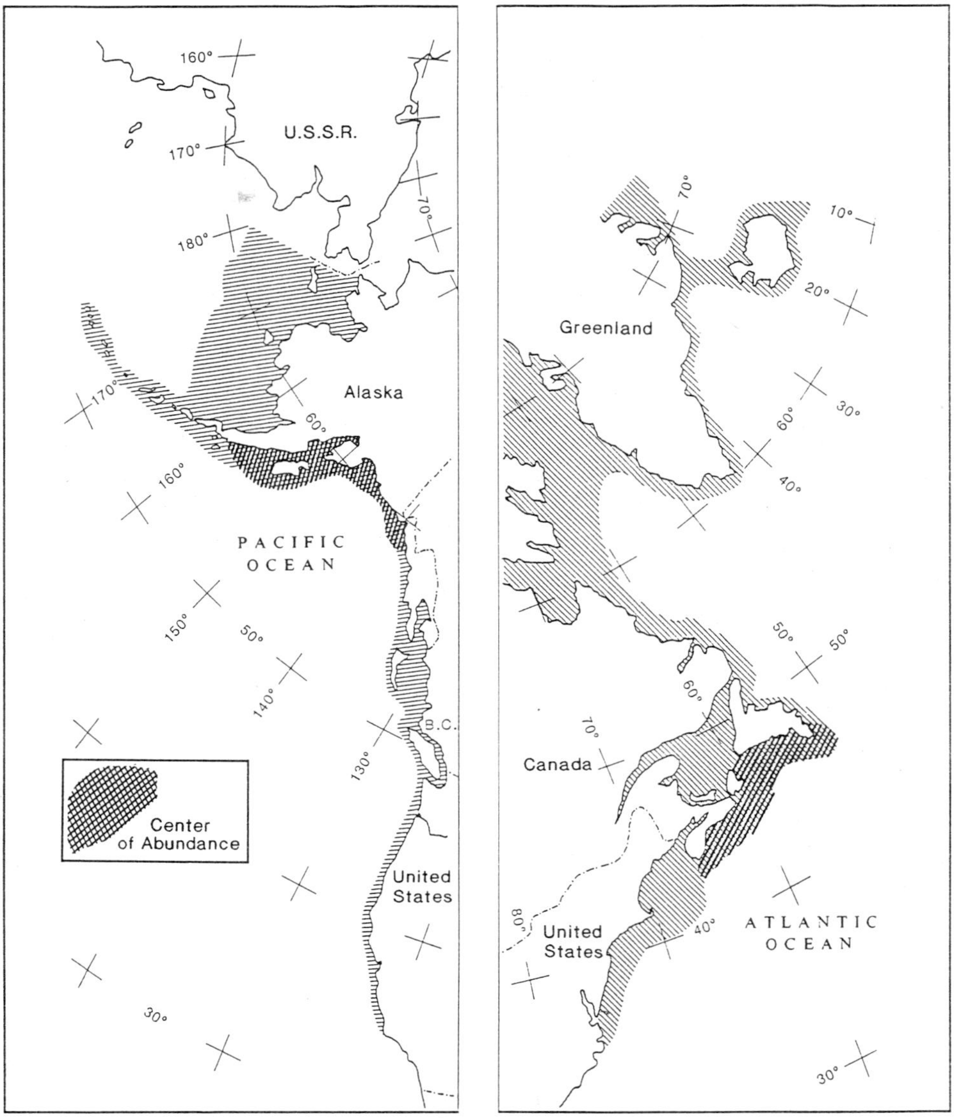

Fig. 4. Distribution of Atlantic and Pacific halibut. *(a)* Eastern north Pacific (after Deriso et al. 1985 and IPHC 1987). *(b)* Western and central north Atlantic (after Scott and Scott 1988 and Haug 1990).

While Grant et al. (1984) offer the strongest support that Pacific and Atlantic halibut differ at the species level, only Vernidub (1936) presents in the scientific literature an objection to separate species. However, rather than try to resolve the controversy with existing data, we recommend completion of future comprehensive studies over the total range of the halibuts to most conclusively settle the issue.

Distribution

North American halibut are typically found along the continental shelf of the sub-Arctic North Atlantic and North Pacific Oceans. Halibut are demersal, living on or near the bottom at temperatures within a few degrees of 5°C. Both species are associated with banks and channels that characterize the continental shelves. The overall latitudinal range of distribution is similar, but the center of abundance is displaced to the south for Atlantic halibut compared with Pacific halibut (Fig. 4). A south-flowing current (the Labrador Current) along eastern North America brings Arctic water that is colder than optimum for halibut onto the continental shelf north of Labrador. A north-flowing current (the Alaska Current) carries warm water along western North America into the Gulf of Alaska at latitudes higher than where comparable temperatures are found in the Atlantic.

Atlantic halibut tend to range slightly deeper in summer than Pacific halibut. Atlantic halibut are most abundant at depths of 200-500 m, although they have been caught within a depth range of less than 100-700 m (Bowering 1986). Summer depth distribution of Pacific halibut is normally less than 200 m, but may extend from 50 to 400 m (IPHC 1987). Larger individuals of both species move to deeper water in winter. In the northwest Atlantic, halibut have been reported as far south as the coast of Virginia to as far north as the Disko Bay area of West Greenland (Fig. 4). They are most abundant in deep water channels running between the fishing banks and along the edge of the continental shelf (Bowering 1986). They are virtually absent from the shallower tops of the fishing banks. This is particularly evident on Saglek Bank and the Newfoundland Grand Bank. In the Newfoundland-Labrador region they are prevalent in the Hudson Strait area, Northeast Newfoundland Shelf, Flemish Pass (the area between the Newfoundland Grand Bank and the Flemish Cap), the southwest and eastern edges of the Newfoundland Grand Bank, and St. Pierre Bank (Bowering 1986). In the Scotian Shelf area, they are distributed along the continental slope and are particularly abundant in the Sable Island Gully area, around Browns Bank and Georges Bank. Inshore concentrations are also known to occur off Cape Sable Island, Grand Manan Island in the mouth of the Bay of Fundy, and in the Minas Basin in the mouth of the Bay of Fundy (Zwanenburg 1984). In the Gulf of St. Lawrence, Atlantic halibut occur in the Esquiman Channel with the area of highest abundance along the north side of Anticosti Island and, to a lesser extent, along western Newfoundland to the northern tip of the channel (McCracken 1958). They also have a tendency to move deeper in winter, particularly larger fish (Bowering 1986). They have been caught in temperature of less than 0°C but are most abundant at temperatures of 5-8°C (Bowering 1986).

The Pacific halibut distribution extends from Santa Barbara, California to Nome, Alaska (IPHC 1987). Adult fish are found in the summer primarily on deep banks at depths greater than 90 m, on the edges of gullies that cut the continental shelf, and on the upper continental slope and outer continental shelf. Pacific halibut are most abundant in the central Gulf of Alaska (Deriso et al. 1985), particularly near Kodiak Island (Fig. 4). High concentrations are also found south of the Alaska

Peninsula and south of the Kenai Peninsula. Moderate abundance is found along both sides of the Aleutian Islands, in the eastern Bering Sea, and through southeast Alaska and northern British Columbia. Limited quantities of adult halibut occur from southern British Columbia to California. In winter, adult halibut move to deeper water on the upper continental slope in the 300-600 m depth range. Pacific halibut are normally found in temperatures from 3 to 9°C (Thompson and Van Cleve 1936).

Reproduction

Sexual determination of live halibut and other flatfish has not previously been possible except through extrusion of spawning products from mature individuals. A procedure for determining the sex of live Pacific halibut from examination of the cloaca, or genital vent, has recently been developed (St-Pierre 1992). The projecting cloaca has a different shape for males and females (Fig. 5). During the first test of the procedure, sex was successfully determined for 101 of 102 Pacific halibut. Subsequently, other IPHC personnel correctly determined the sex of 97% of 869 Pacific halibut. The procedure is most suitable for individuals longer than 50 cm, as the cloaca is too small for successful sex determination in smaller fish. The method

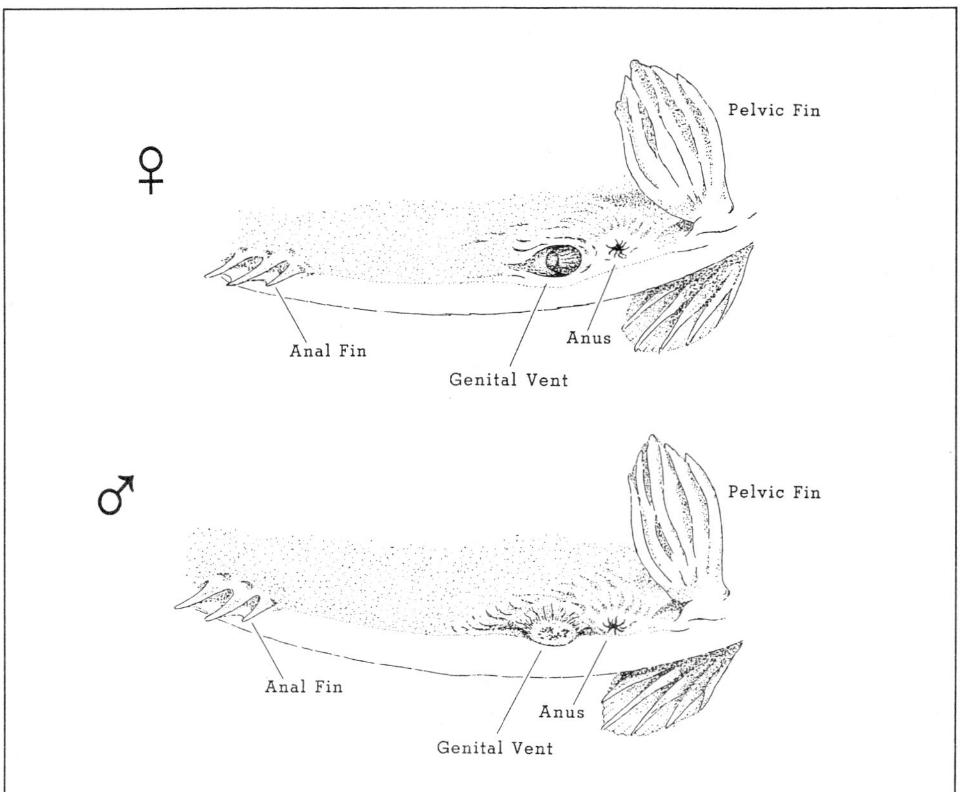

Fig. 5. Ventral view of female and male genital vents in Pacific halibut (from St-Pierre 1992).

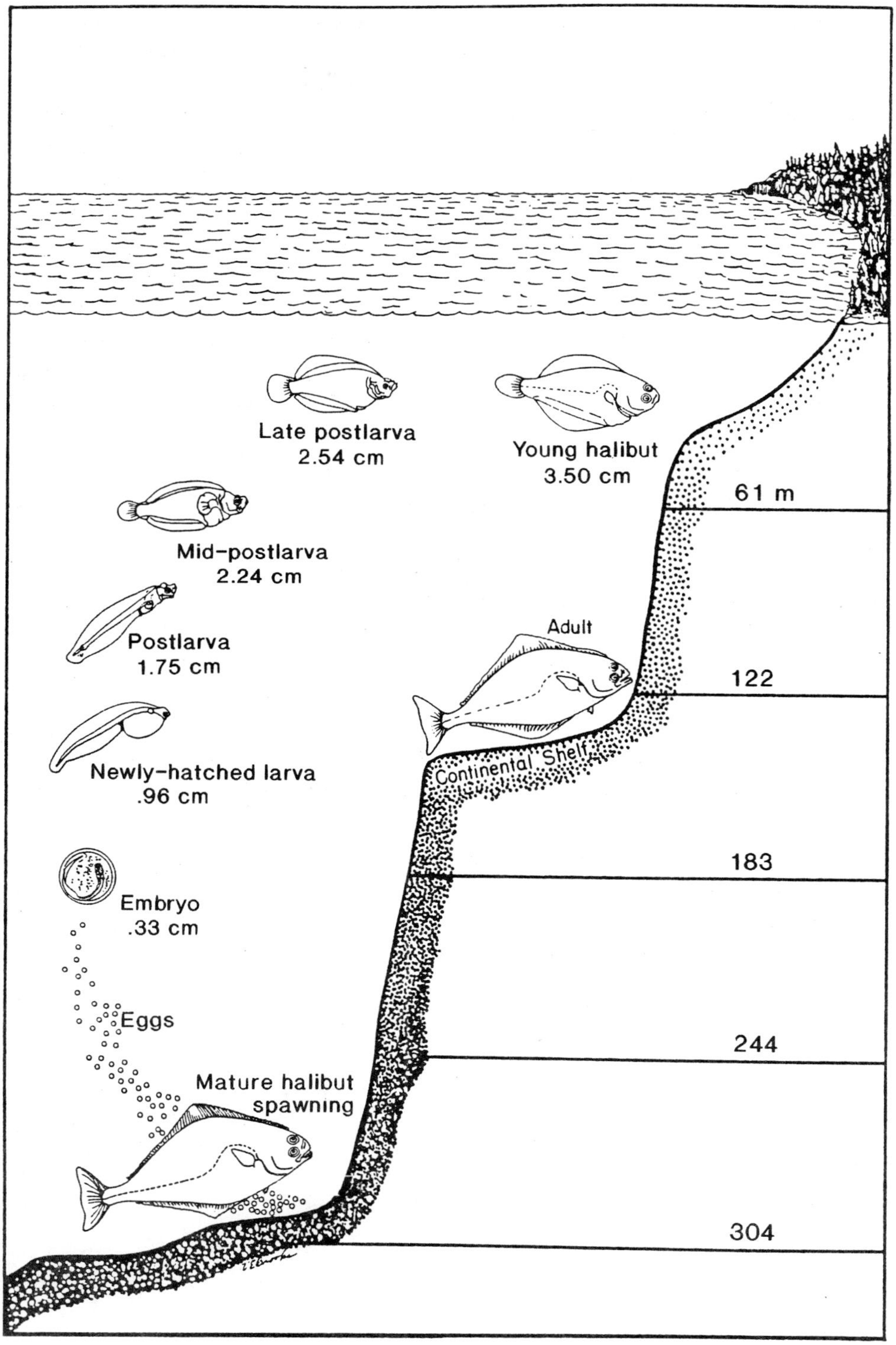

Fig. 6. Schematic life cycle for Pacific halibut (from IPHC 1987), probably applicable also for Atlantic halibut.

may have application for other flatfish species, as IPHC staff have also been successful in differentiating the sex of arrowtooth flounder (*Atheresthes stomias*).

Schmitt and Skud (1978) and Bowering (1986) concluded that the onset of maturity is strongly related to size of fish. This conclusion agrees with a general life history pattern of fishes (Bagenal 1978). The importance of fish age in many biological and stock assessment analyses often leads to the examination of fecundity-age relationships. Because size-at-age changes if growth rates change, fecundity-size relationships are more stable than fecundity-age relationships.

Atlantic and Pacific halibut reproduce in a nearly identical manner (St-Pierre 1984). Males and females broadcast eggs and sperm, and fertilization occurs externally by random contact (Fig. 6). Spawning grounds for Pacific halibut (Fig. 7) were identified during the years when fishing occurred during the spawning season (St-Pierre 1984). Very little information is available for spawning locations of North American Atlantic halibut, although a study is underway to address this deficiency.

In general, small, first-spawning halibut may release only 100 000 eggs, while a large female may release over 4 million eggs. Schmitt and Skud (1978) fitted a power curve to Pacific halibut fecundity relationships and found significant differences for halibut from the Gulf of Alaska and those taken from waters off British Columbia. Schmitt and Skud (1978) showed that fecundity is more closely related to fish size than to age. They also found major differences in fecundity-at-age, which more than doubled in the Gulf of Alaska from 1927 to 1973 because growth rates increased during this period.

Males and females mature over a range of ages that varies somewhat by area (Novikov 1964; Schmitt and Skud 1978; St-Pierre 1984). Novikov (1964) indicated that Bering sea halibut mature at ages 7-13 years for males and 9-15 years for females, Gulf of Alaska halibut mature at 5-11 years for males and 8-16 years for females, and Norwegian halibut mature at 7-17 years for males and 8-18 years for females. St-Pierre (1984) found similar ages for Pacific halibut.

Schmitt and Skud (1978) pointed out that estimated mean age of sexual maturity will change with season: mean age will be lower as mature fish migrate to winter spawning grounds, because immature fish of a given age will be less available. They found the age of 50% maturity for female Pacific halibut in the eastern Gulf of Alaska from 1954 to 1957 to range from 8.8 to 10.7 years in winter and from 11.8 to 12.4 years in summer. Schmitt and Skud (1978) also found that age of maturity and fecundity changed over time for females in the Gulf of Alaska; during the 1949-1959 period, the age of 50% maturity was 11.8 years, but 10.9 years from 1960 to 1977. A similar change did not occur in waters off southeast Alaska and British Columbia, for which age of 50% maturity declined only from 12.4 to 12.1 years. Schmitt and Skud (1978) attributed difference in maturity by area to an increased growth rate in the Gulf of Alaska that did not occur in southeast Alaska and British Columbia.

St-Pierre (1984) presented percent maturity by age and sex for Pacific halibut caught during setline research surveys in Alaska waters, and Bowering (1986) examined the size and age of sexual maturation of Atlantic halibut caught during research vessel surveys (bottom trawl) off southern Newfoundland. Both analyses have been updated using probit analysis (Trippel and Harvey 1991) for the purpose of this review. Atlantic halibut are reported to mature first as 4 and 6 year olds, for males and females, respectively. Pacific halibut first mature at ages 5 to 6 and 7 to 8 for males and females, respectively. The length at which 50% were mature (M_{50}) is similar for female Atlantic and Pacific halibut, at about 119 cm (Fig. 8). Atlantic and Pacific males, however, appear to reach maturity at different rates. Gilbert St-Pierre (IPHC, Seattle, Wash., personal communication) cautions that some Pacific halibut maturity data for males may have been compromised by inexperienced biologists misreporting females as males.

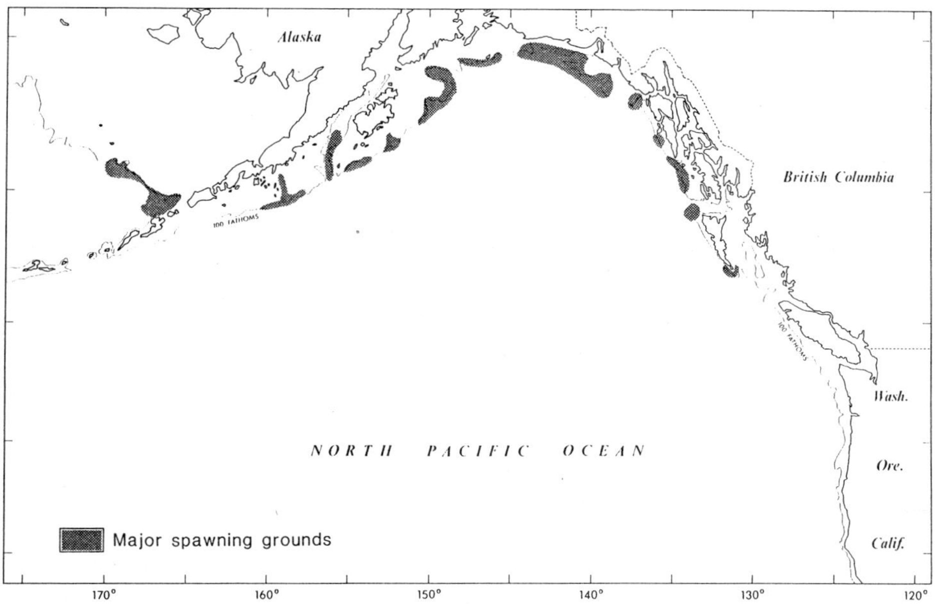

Fig. 7. Major spawning grounds for Pacific halibut in the eastern north Pacific (from St-Pierre 1984).

Results provided here for females are similar to those obtained by McCracken (1958) in an investigation of Scotian Shelf halibut captured with longline gear. However, the length at which 50% of males were mature was considerably greater than our values, possibly resulting from the comparatively small sample size available to McCracken. Alternatively, such differences may reflect general declines in median size at sexual maturity noted over a similar period for other flatfish species (Beacham 1983*a*,1983*b*).

Spawning season is similar for Atlantic and Pacific halibut, occurring mainly in late fall and winter for Pacific and in winter and spring for Atlantic halibut (Novikov 1964; St-Pierre 1984; Jákupsstovu and Haug 1988). November through March is the normal spawning period for Pacific halibut. Atlantic halibut spawning extends from December to April in Norwegian waters, from March to May off Iceland, and from February to April in the western Atlantic (references in St-Pierre 1984; Jákupsstovu and Haug 1988).

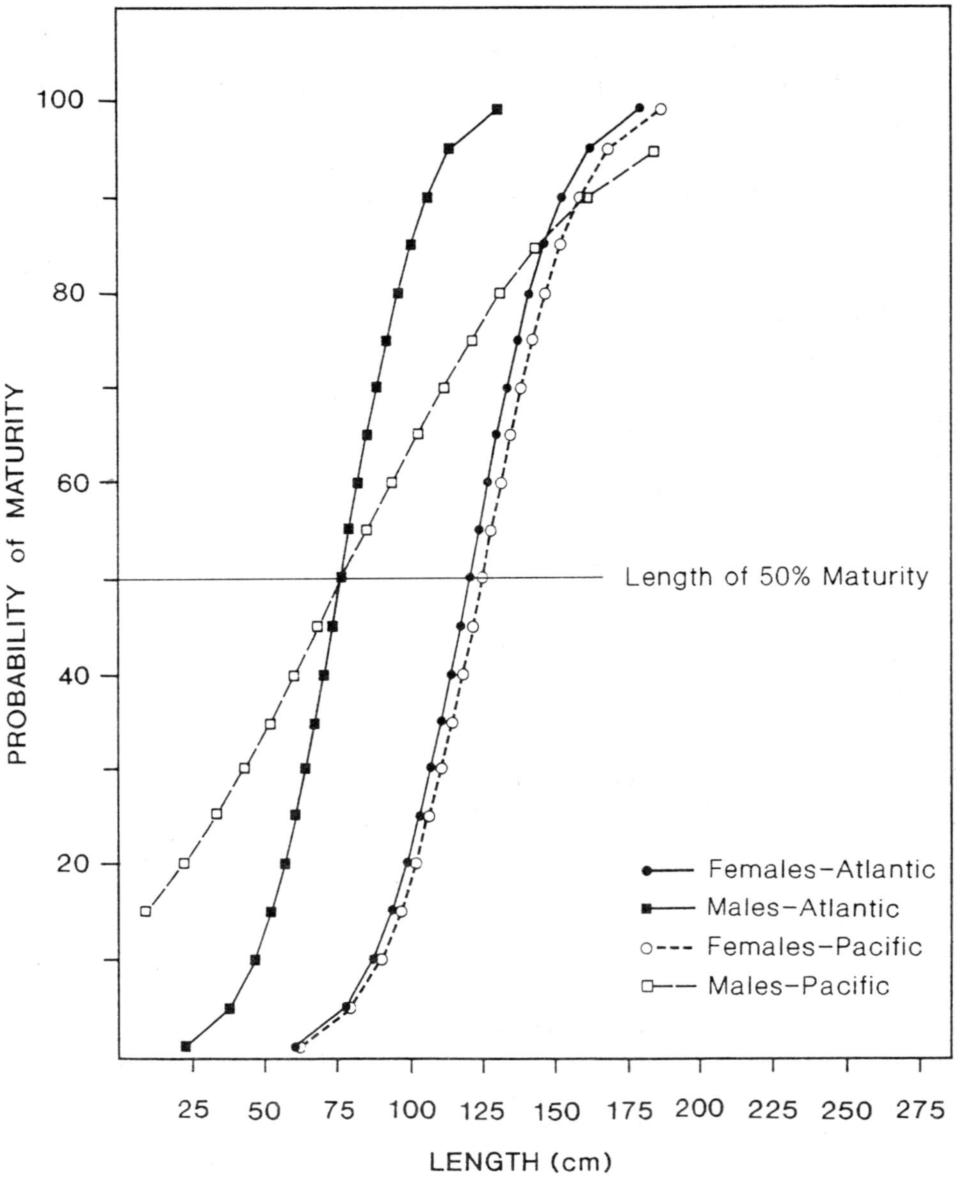

Fig. 8. Maturity ogives for male and female Atlantic and Pacific halibut. Atlantic halibut captured during research cruises in the Canadian northwest Atlantic; Pacific halibut captured during research cruises in the northeast Pacific.

Thompson (1915, 1916) and Kolloen (1934) concluded that mature Pacific halibut ovaries contain several generations of ova that mature from one year to the next, and observed small batches of maturing eggs from the current generation. Their conclusion is supported by a micrograph of Atlantic halibut gonads (Fig. 9).

Spent ovaries contained a mass of newly differentiated ova that will develop the next year, and undifferentiated oocytes that represent future spawning. The pattern of oocyte development in halibut fits that described by Wallace and Selman

(1981) for group-synchronous spawners, with distinct classes of oocytes apparent in the ovary at most stages of maturation (Neilson et al. 1993).

The group-synchronous pattern is the most common type of ovarian physiology among teleosts. Oocyte development follows the general pattern of the development of multiple nucleoli which eventually disperse throughout the ooplasm. The next stage entails the sequestering from the maternal blood supply of the yolk precursor (vitellogenin), which is packaged into yolk vessicles (Fig. 9). The latter generally fuse during oocyte maturation forming a continuous fluid phase surrounded by a peripheral layer of cytoplasm. The final stage of development of the oocyte prior to release into the lumen is hydration. The large number of annually maturing eggs requires batch spawning over a period of time, as a female would be unable to hold all fully developed eggs at one time. The actual duration of spawning by an individual fish is unknown.

Spermatogenesis in the halibut also seems to fit the teleost norm (J. D. Neilson, St. Andrews Biological Station, St. Andrews N.B., unpublished data), with spermatogonia occurring along the length of the tubular structure of the testes (Fig. 9), as compared with atheriniform fish where spermatogenesis is confined to the terminus of the tubule only (Grier 1981).

Thompson (1915) concluded that spawning is an annual event. Novikov (1964), however, presented information suggesting that halibut do not spawn every year. He observed halibut at the end of the spawning season (February and March) with developing ova presumably too immature to spawn that year. St-Pierre (1984) noted that Novikov's (1964) observation may have been of immature halibut that had not yet begun to spawn, because a number of female halibut will not spawn for the first time until 16 years of age. St-Pierre (1984) reported that IPHC research data during the spawning season indicated that over 94% of 16-year-old female halibut were in various stages of spawning; roughly 50% should not be in spawning condition if the fish spawn only every other year. Therefore, we conclude that halibut are annual spawners.

Early Life History

The pelagic phase of existence for both halibuts is thought to last 6-7 months. While there have been several studies of Pacific halibut early life history, comparatively few data on the pelagic life history stage are available for Atlantic halibut. Spawning occurs at the edge of the continental shelf in areas with soft clay or mud bottom at depths of 300-700 m for Atlantic halibut (Devold 1938), and near the edge of the continental shelf, generally at depths between 180-550 m for Pacific halibut (St-Pierre 1984) (Figs. 6 and 7). The general egg morphology of both halibuts is similar. The egg contains a large yolk and small perivitelline space (Thompson and Van Cleve 1936; Lønning et al. 1982). A pattern of "stellate cross hatching" described for Atlantic halibut by McIntosh (1892, reference in Thompson and Van Cleve 1936) was not evident in the egg samples examined by Thompson and Van Cleve, who described the surface of the egg shell as clear, glass-like and minutely

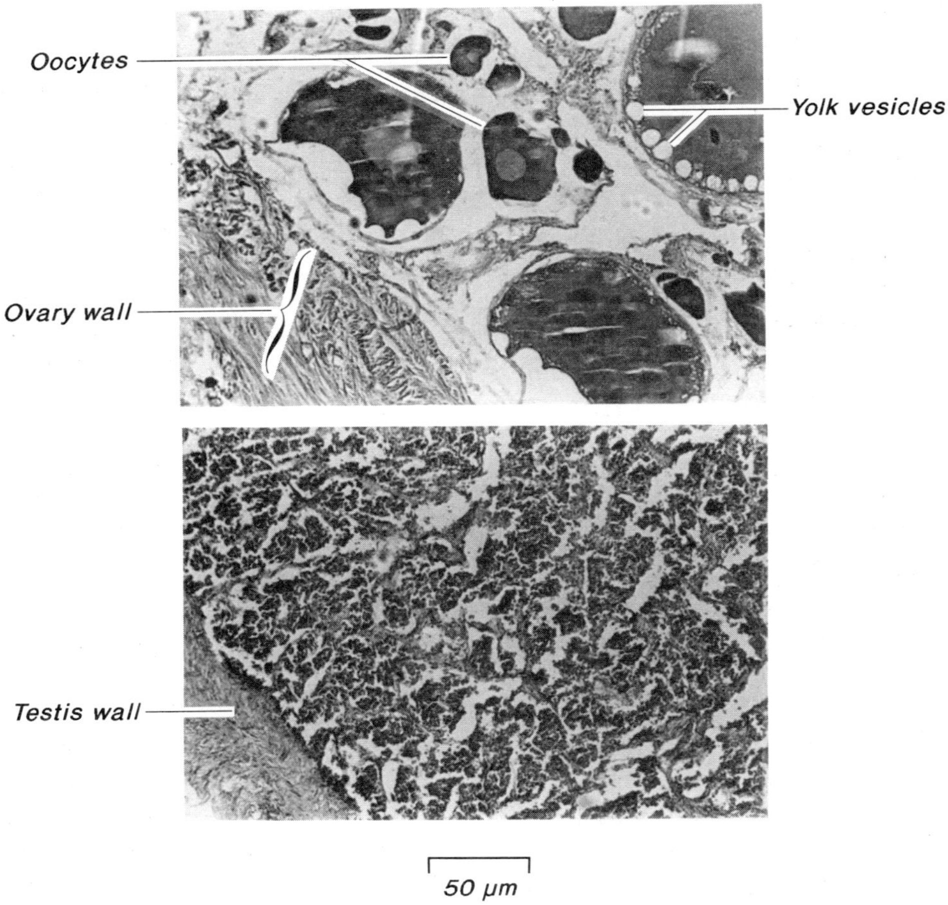

Fig. 9. Micrographs of Atlantic halibut gonads stained with Harris' haematoxylin and counterstained with eosin: *(a)* ovary; *(b)* testis.

honeycombed. Lønning et al. (1982) noted that the chorion of Atlantic halibut is characterized by a greater number of lamellae (18) than other flatfish species (6-9), and that the morphology, ultrastructure, and development resembles patterns found in other pelagic eggs. Such comparisons have not been made for Pacific halibut.

The eggs of both halibuts are exceptionally large by marine teleost standards, with a mean diameter of fertilized eggs before first cleavage of about 3 mm. In north Norway, egg diameter varied between 2.86 and 2.98 mm, and averaged 2.92 mm (Lønning et al. 1982). Thompson and Van Cleve (1936) determined a size range of 2.90-3.80 mm, averaging 3.27 mm, for 2824 Pacific halibut ova collected from plankton samples, and a range of 2.05-3.85 mm, averaging 3.17 mm, for eggs taken from a live spawning female. They attributed the smaller size of eggs from the live fish to the presence of immature eggs. Mean diameters of fertilized Pacific halibut eggs were reported as 3.2 and 3.34 mm by Forrester and Alderdice (1973) and 3.4 ± 0.9 mm by McFarlane et al. (1991). Egg diameter of Atlantic halibut tends to decrease during the spawning season (Kjørsvik et al. 1987), but comparable data are not available for Pacific halibut. In general, field and laboratory studies on both Atlantic and Pacific halibut eggs suggest that they are initially bathypelagic. Thompson and Van Cleve (1936) and Van Cleve and Seymour (1953) for Pacific halibut and Haug et al. (1986) for Atlantic halibut caught newly spawned eggs with plankton nets, and were able to determine the vertical distribution of the eggs. Van Cleve and Seymour (1953) found concentrations of Pacific halibut eggs more closely aligned to density than to depth. Haug et al. (1986) showed that the vertical distribution of halibut eggs was associated with the prevalent hydrography, with most dense concentrations often found at or just above the pycnocline under stratified conditions.

Most laboratory studies confirm that neutral density of eggs occurs at bathypelagic depths (Forrester and Alderdice 1973; Haug et al. 1984, 1986; and McFarlane et al. 1991) although Lønning et al. (1982) concluded that Atlantic halibut eggs are buoyant at 37‰ at ambient temperature, considerably higher than salinities found on the northeastern (Lønning et al. 1982) or northwestern (Drinkwater and Trites 1987) continental shelves of the North Atlantic Ocean. However, these reports do not agree on changes in egg density with increased development. Haug et al. (1984, 1986) found Atlantic halibut eggs buoyant at salinities of 33.8-35.0‰ in the eastern Atlantic, and indicated that Atlantic halibut eggs rise for 2-4 days after spawning deposition to a depth of neutral density, but that later stage eggs are neutrally buoyant at higher densities (as indicated by salinity). Forrester and Alderdice (1973) also found the salinity of neutral buoyancy (at constant temperature) to increase with time from hatching. However, McFarlane et al. (1991) measured salinity of neutral buoyancy for Pacific halibut in the laboratory and estimated location of developing eggs in the water column. They predicted spawning depths of 200-600 m, and a rise of fertilized eggs to 100-200 m depth within 12 to 18 days.

Hatching of Atlantic halibut occurs after 18 days of incubation at 5°C, at what is considered to be an early stage of development (no functional eyes or mouth) relative to other teleosts (Lønning et al. 1982). In contrast, Forrester and Alderdice (1973) found that Pacific halibut took 21 days to hatch at 5°C and 12.5 days to hatch at 8°C. Similarly, McFarlane et al. (1991) found hatching in 18 days at 5°C and 13 days at 7°C. Forrester and Alderdice (1973) did not find hatching at temperatures lower than 5°C, but Lønning et al. (1982) observed hatching at 2.5°C, albeit with reduced survival. The yolk sac contains a large reserve, as Øiestad and Haugen (1980) found that absorption of the yolk was not complete until 1.5-2.0 months after hatching.

Data on larvae are typically scarce, probably because they are difficult to catch in sufficient numbers. Length at first hatch has been measured by several workers for Atlantic and Pacific halibut, but with contrasting results. Thompson and Van Cleve (1936) first noted a hatching length of 8-15 mm for Pacific halibut. Forrester and Alderdice (1973), who also worked with Pacific halibut, measured length at hatching of about 6 mm, ranging from 5.33 to 7.79 mm. McFarlane et al. (1991) obtained an intermediate value of 7.5 $\pm$ 0.21 mm. Rearing experiments on Atlantic halibut, indicate that the larvae are about 6-7 mm long at hatch (Lønning et al. 1982; Haug 1990), similar to the results of Forrester and Alderdice. Pittman et al.

(1987), working with Atlantic halibut, noted that eyes were fully pigmented at around 180 degree-days, as was the jaw apparatus. On the basis of the time of full absorption of the yolk, those authors suggested that halibut larvae were fully adapted to exogenous feeding after about 250-300 degree-days. McFarlane et al. (1991) observed hatching of Pacific halibut at 7-8 mm, and monitored growth and development until the last larva died at 13.5 mm, 56 days post-hatch.

The long pelagic phase of eggs and larvae for both halibuts provides a mechanism for long distance redistribution from the spawning grounds. The drift of Pacific halibut eggs and larvae have been extensively examined by the IPHC. Skud (1977) reexamined ichthyoplankton data of Van Cleve and Seymour (1953) and concluded that a definite northerly movement of eggs occurred from prevailing surface and subsurface ocean currents, which carry them from spawning grounds near Cape St. James (Queen Charlotte Islands) as far north as Alaskan waters. His analyses caused a reconsideration of the view held previously that halibut stocks off British Columbia and Alaska waters were separate and independent. The current pattern in

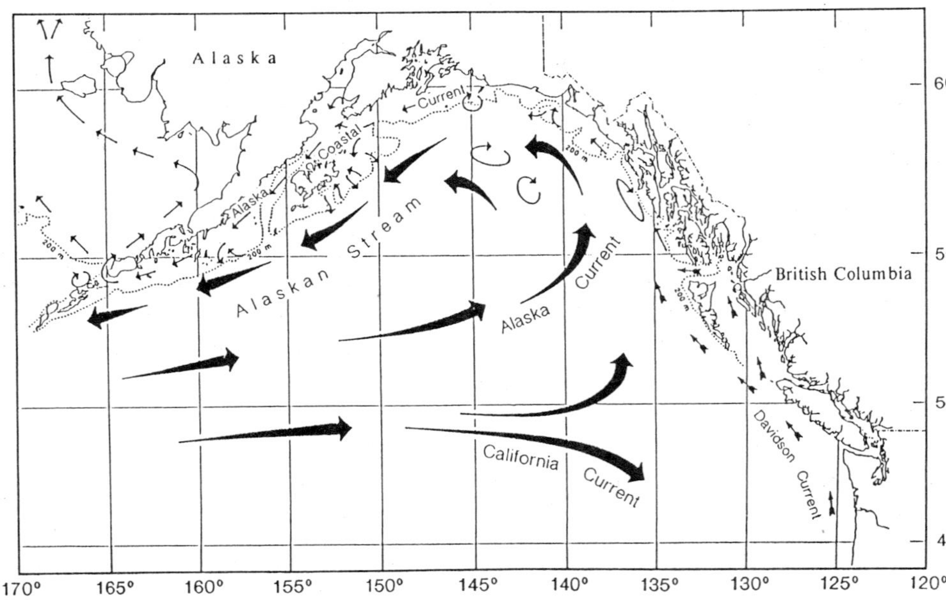

Fig. 10. Surface currents in the eastern north Pacific during the drift of Pacific halibut eggs and larvae.

the eastern North Pacific during winter (Fig. 10) shows the probable path of egg and larval drift, and St-Pierre (1989) has provided supporting evidence.

Comparatively few data on the distribution of early life history stages of Atlantic halibut are available, particularly for the northwest Atlantic stock. Cox (1924) found two specimens relatively close inshore and speculated that an investigation of the continental slope off Nova Scotia would lead to the discovery of halibut eggs. However, in the course of extensive ichthyoplankton programs covering the Scotian Shelf (Scotian Shelf Ichthyoplankton Program [1978-82] and the Fisheries Ecology Program [1983-84]), very few halibut eggs or larvae were found. The majority of ichthyoplankton stations were completed in waters less than 200 m depth. Results from depth-stratified annual research surveys suggest that Browns Bank may be a significant rearing area for young Atlantic halibut because individuals in the stratum corresponding to Browns Bank were consistently smaller than those from other strata (Neilson et al. 1993).

Atlantic and Pacific halibut settle to the bottom at different stages of the larval phase. According to Tåning (1936), Atlantic halibut larvae settle to the bottom and adopt a demersal habit before completing pigmentation, eye movement, and fin formation. The metamorphosis to a typical adult shape occurs at lengths of 34-44 mm. Pacific halibut are still in the water column after fin formation, after the eye has moved past the dorsal profile, and after pigmentation is nearly complete (Thompson and Van Cleve 1936). Haug and Sundby (1987) note that nursery areas are apparently well defined and are often shallow (20-60 m) coastal areas with sandy bottoms. Examples of such nurseries for Atlantic halibut include the Faroe Islands (Tåning 1936), Faxa Bay on the west coast of Iceland (Sigurdsson 1956), and possibly in the vicinity of Sable Island Gully off Nova Scotia (Stobo et al. 1988). In the Pacific, halibut larvae are found in the water column at depths of 180 m or less by the age of 3-5 months. They are then carried inshore by the currents, where they settle to the bottom in shallow waters in late spring and early summer (St-Pierre 1989). The distribution of the juvenile halibut seems quite dispersed as they occur over the entire continental shelf from Cape Spencer to Unimak Pass, and in the Bering Sea (Best 1968). However, some discrete concentrations have been identified, such as the ones that occur on Portlock Bank (Skud 1977).

Movements of Juvenile Halibut

Continuity of the Pacific halibut resource requires that the progeny move back to the east and south at some stage in the life history to counter the extensive drift of eggs and larvae. Best (1968) and Skud (1977) present evidence that the counter migration for Pacific halibut occurs during the juvenile stage, and that most juveniles migrate between ages 2 and 6 years. There is insufficient information on ichthyoplankton distribution and spawning sites for Atlantic halibut to infer compensatory movements of juveniles. Stobo et al. (1988) showed that juvenile (<75 cm) Atlantic halibut in the northwest Atlantic exhibit more extensive migrations than do adults, consistent with findings for Pacific halibut. However, the movements of Atlantic juveniles appear to be non-directed (Haug and Sundby 1987; Godø and Haug 1988).

Recovery data from juvenile Pacific halibut tagged from trawlers as 2, 3, and 4 years old show considerable migration to the east and south (Fig. 11). Tagged juvenile halibut have rarely been documented moving to the west or from the Gulf of Alaska into the Bering Sea. Extreme migration from the Bering Sea to waters off northern California has been documented (IPHC, Seattle, Wash., unpublished data), but migration from the Bering Sea, Unimak Pass, and Kodiak Island areas to southeast

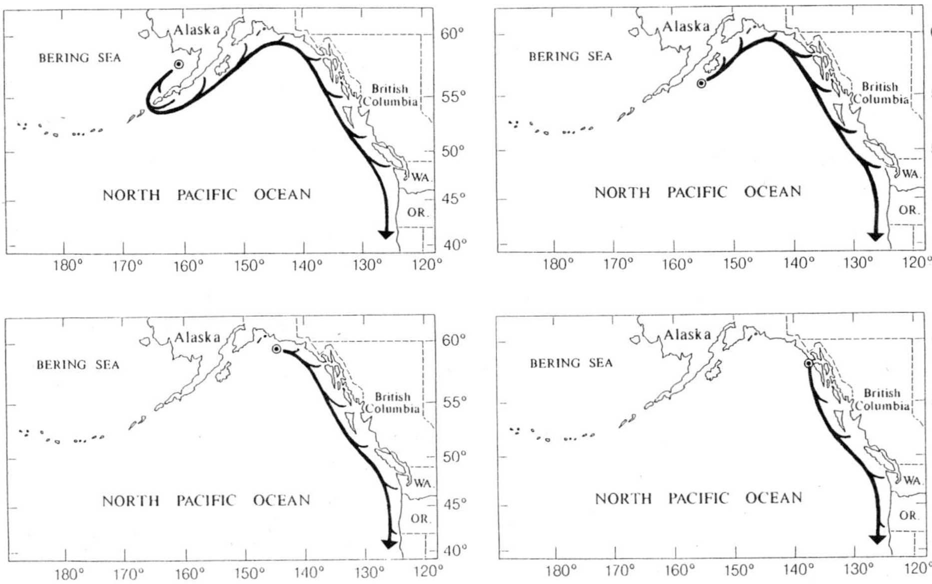

Fig. 11. Schematic movements of juvenile Pacific halibut (<65 cm in length) as determined from tagging experiments.

Alaska and British Columbia is consistent. Migration rates cannot be reliably calculated from available data because recovered tags are not consistently returned from different regions of the coast (Trumble et al. 1990).

Age composition of migrating juvenile Pacific halibut presented by Skud (1977) and updated in Fig. 12 (IPHC, Seattle, Wash., unpublished) supplies additional information supporting countermigration, primarily by juvenile halibut. One- and 2-year-old Pacific halibut are commonly found by trawl surveys in inshore areas of central and western Alaska, but are virtually missing from southeast Alaska and British Columbia. Juvenile halibut tend to move further offshore at 2 or 3 years of age. Mean and modal ages in the offshore surveys increase to the east and south, with the youngest trawl-caught halibut off southeast Alaska and British Columbia tending to be 4 years of age and older.

Northeast Atlantic populations described by Haug and Sundby (1987) and Godø and Haug (1988) initiate movements away from the nursery grounds at about 3 years of age. Joensen (1954) has suggested that faster-growing females might leave the nursery grounds prior to the males.

Movements of Adult Halibut

By the time Pacific halibut recruit to the commercial fishery (at a minimum size of 81.3 cm), most of the extensive countermigration to balance egg and larval drift has apparently taken place. Commercial sized halibut, however, continue seasonal movements but with very little directed migration from year to year.

Early analyses of tagging data by the IPHC suggested that mature halibut did undertake large directed migrations. Thompson and Herrington (1930) initiated

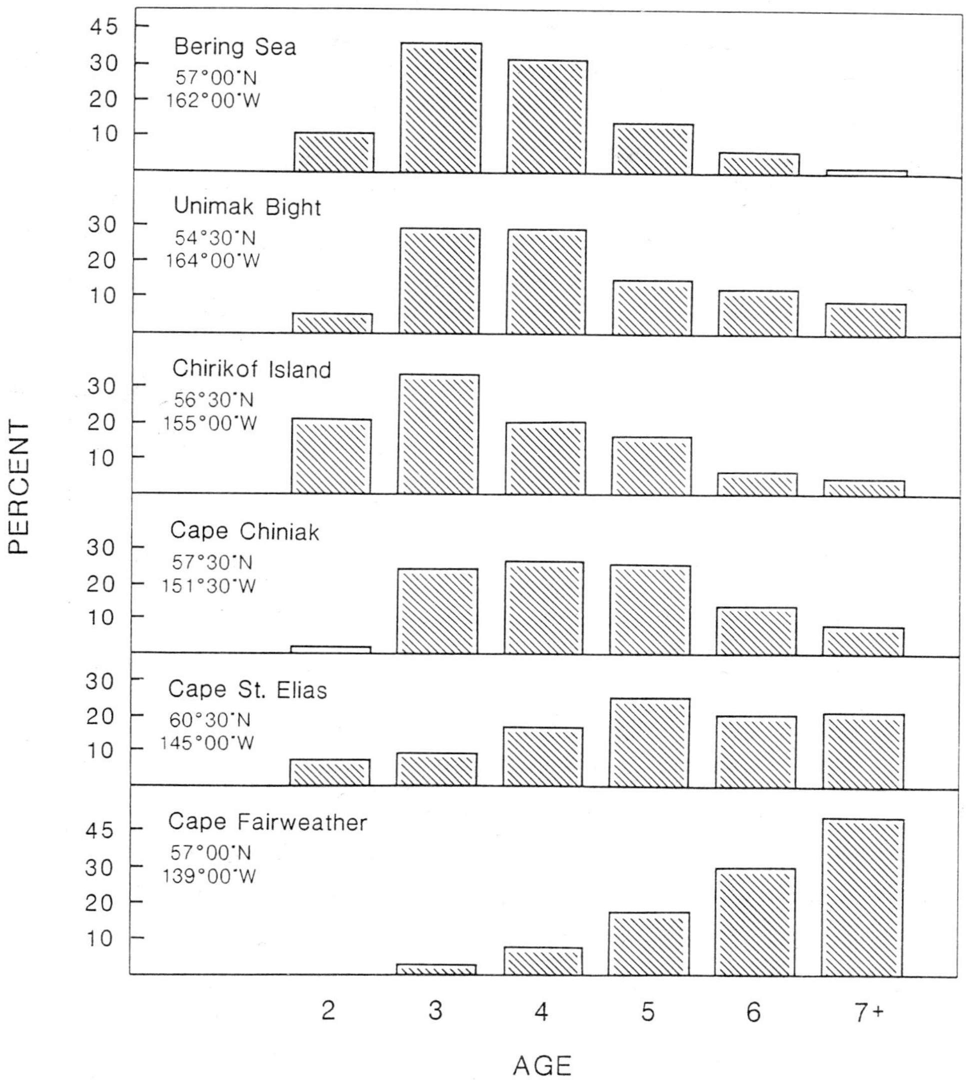

Fig. 12. Distribution of catch in percent numbers of Pacific halibut by age in coastal areas, 1973–1980.

tagging in 1925, often in winter, over the range of fishing activity (Vancouver Island to the Alaska Peninsula), so that both releases and recoveries would be representative of any populations harvested. They concluded that separate populations existed north and south of Cape Spencer, Alaska. As a result, the IPHC created separate management areas. Tag recoveries were interpreted to show 400-500 km westward migrations by predominately mature fish in the Gulf of Alaska but limited movements by predominately younger fish in waters of southeast Alaska and British Columbia.

Skud (1977) demonstrated that analysis of early tagging experiments did not recognize the seasonality of Pacific halibut migration. Most tagging in the Gulf of Alaska originally occurred during winter, but recoveries were made during the

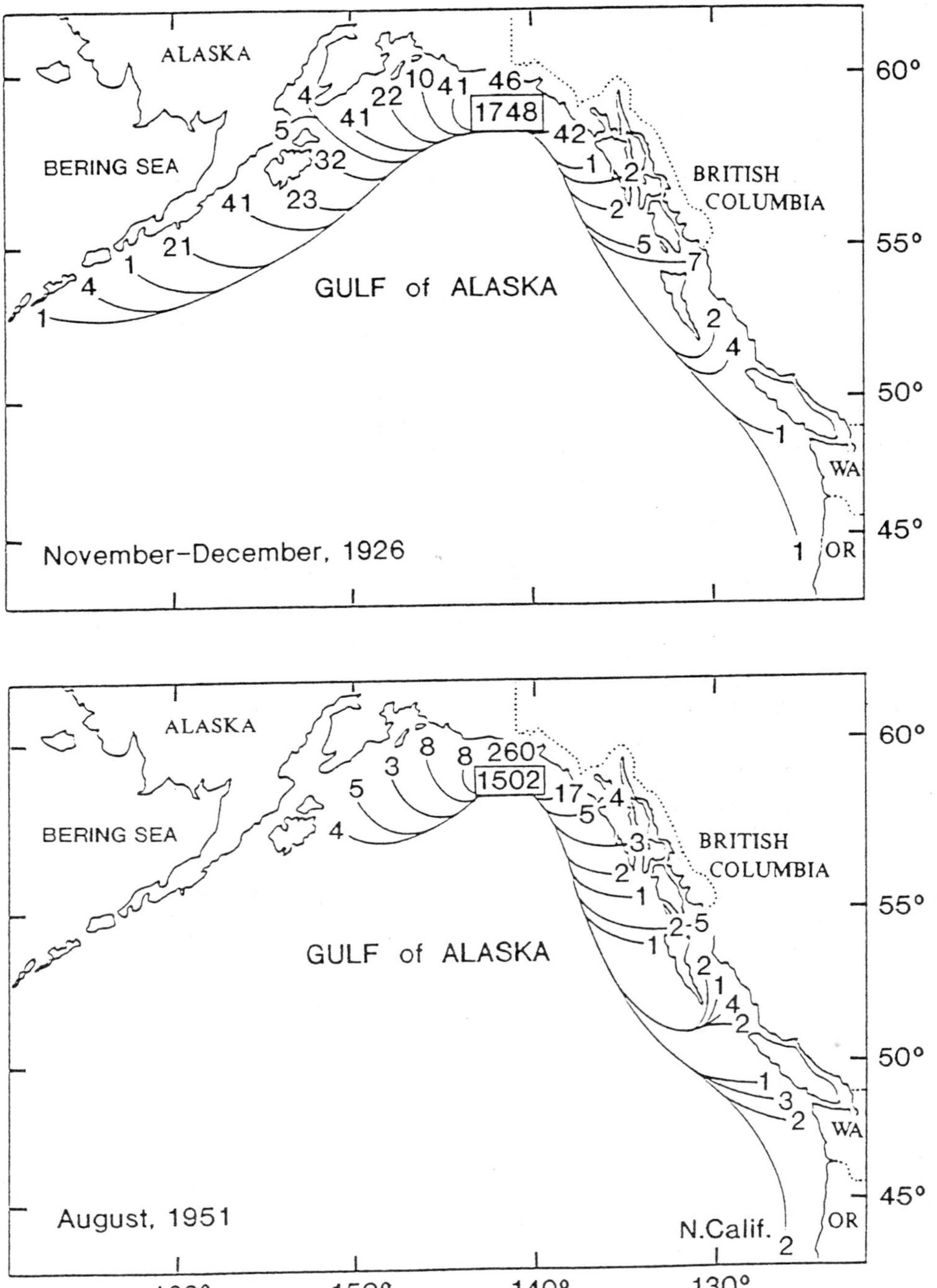

Fig. 13. Distribution of recoveries from Pacific halibut tagged off Yakutat in November–December 1926 and in August 1951. The number of tagged fish is shown in the box (from Skud 1977).

February to November fishing season; conversely, most tagging and recoveries off southeast Alaska and British Columbia occurred during summer. Skud (1977) examined data from several seasons to show that mature Pacific halibut migration is limited predominately to movements from summer feeding grounds during the autumn to winter spawning grounds and return during spring.

Once a summer feeding location is established, mature halibut show little net summer to summer movement. Skud (1975) noted that emigration was observed from all areas of tagging and that several legal sized Pacific halibut have migrated over 1600-2000 km from the Bering Sea to Washington state or south. Most of the fish tagged and recovered in summer, however, were recovered in the area of release. Conversely, summer recoveries from winter releases or winter recoveries from summer releases tend to be widespread from the release site. The distant movements of Pacific halibut examined by Thompson and Herrington (1930) proved to be mostly winter to summer movements. Skud gave examples of the amount and distance of seasonal migrations. Summer tagging in 1954 in the eastern Gulf of Alaska had 83% of the tags recovered in the area of tagging, with no recoveries more that 500 km to the west. Winter tagging in 1955 in the same vicinity had only 63% of tags recovered in the area of tagging, with more than 10% of recoveries beyond 500 km west. Similar results were obtained from northern British Columbia where a 1947 summer release had 96% of recoveries in the area of release and only 2% beyond 200 km, while a 1940 winter release had 33% recovered in the area of release and 8% beyond 200 km. Figure 13 shows the dispersion from winter tagging and the central tendency for summer releases. Most winter 1926 tags were recovered out of the tagging area, and only 21% of them moved to south and east locations. Most of the 1951 summer tags were recovered in the area of tagging, but 67% of those that were not, moved south and east.

Commercial fishing for Pacific halibut currently occurs only in late spring to autumn, so summer to summer movement patterns are of most interest for management purposes. A large scale, multipurpose summer longline survey released tagged fish on a grid since 1976 with the expectation that migration rates would be determined from tag recoveries. Deriso and Quinn (1983) and Quinn et al. (1985) estimated migration rates for exploited fish from the grid and earlier surveys. They found that few commercial sized Pacific halibut migrated from the management area of release, although emigration rates from the Bering Sea were higher than from the Gulf of Alaska. The actual migration rates are uncertain, however, because tags are not returned at a consistent rate by time and area (Trumble et al. 1990).

Northeast Atlantic adult halibut show similar patterns as do those on the Pacific coast, as Godø and Haug (1988) note that mature fish leave the spawning grounds when the spawning season is over and disperse in a random fashion, apparently in search of food. Godø and Haug (1988) further noted that results from Norwegian tagging experiments, where adult halibut were tagged and released on spawning grounds, showed that recaptures 1-3 years later were either at tagging sites (if recaptured during spawning season) or outside the tagging sites (if recaptured during a period other than spawning).

The movements of northwest Atlantic halibut were reviewed by Stobo et al. (1988). Martin and McCracken (1950) reported on tagging studies conducted in the Gulf of St. Lawrence and on the southwestern portion of the Scotian Shelf between LaHave and German Banks. In the Gulf of St. Lawrence, 76 animals were tagged in June 1946 off Anticosti Island; 8 had been recovered when the authors reported their findings. Of these, five had moved only 3-5 km from their point of release. However, a larger animal had travelled more than 160 km in only 1 month. The southwestern Nova Scotia study involved 154 halibut tagged in 1947, of which 16 had been recovered at the time of Martin and McCracken's (1950) publication. Recoveries were to the east of the tagging site, with two halibut recovered over 300 km away on

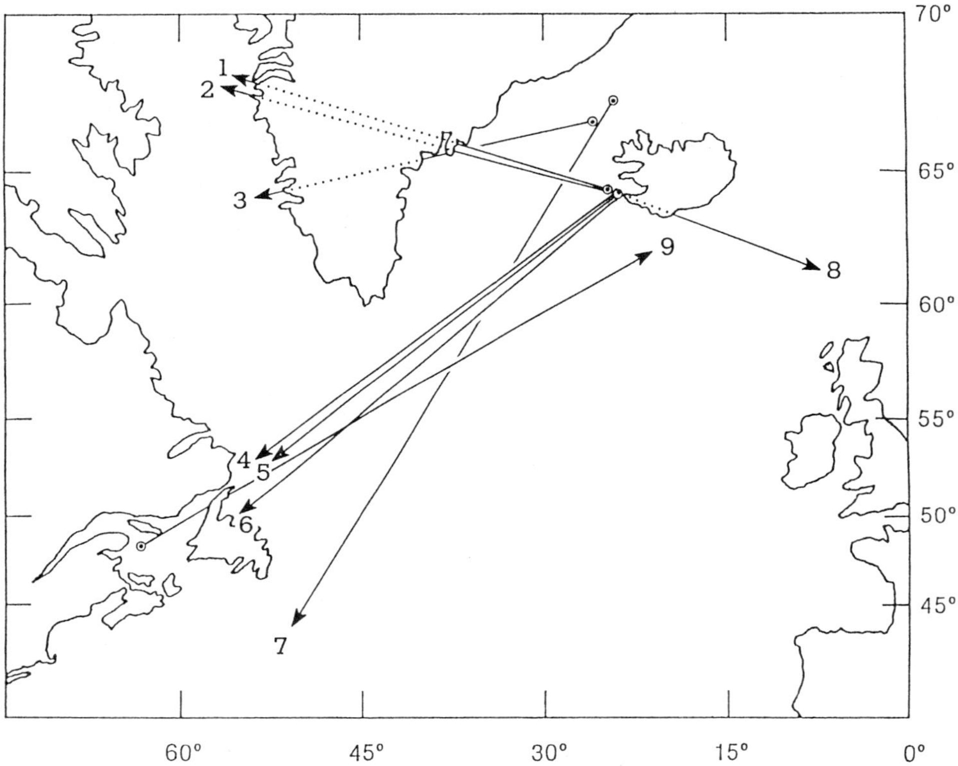

Fig. 14. Examples of long-distance migrations of Atlantic halibut. Data on fish Nos. 1–8 were provided courtesy of Dr. G. Jonsson, Hafrannsoknastofnunin Marine Research Institute, Iceland, and data on fish No. 9 were from McCracken (1958).

Middle Ground and the Western Bank. Most of the recoveries, however, showed movements over short distances only.

McCracken and Martin (1955) updated their previous marking studies in the Gulf of St. Lawrence and on the southwestern portion of the Scotian Shelf, with a further seven recoveries made during the period January 1950 to October 1954. Six of the seven fish originated from the southwestern Nova Scotia tagging. The movements inferred from these additional data were also to the east, similar to their previous report. Of the two more distant recaptures, one was taken off Sable Island in 1952 and the other on the southern edge of the Grand Bank in 1953, a straight line distance of about 960 km. Of special note was the single recapture from the Gulf tagging experiment, which was taken near west Iceland, a straight line distance of about 2560 km (Fig. 14, fish No. 9). Other examples of long-distance migrations are fish marked off western Iceland moving to the vicinity of Newfoundland (Fig. 14, fish Nos. 4, 5, 6, and 7). McCracken and Martin (1955) concluded that halibut may move great distances to widely separated fishing banks, but the majority remain in the area of tagging. They considered such results to be consistent with observations of locally discrete variations in coloration, growth, and size composition, characteristic of a number of populations. However, no source or reference was given with regard to the local variations in phenotype.

McCracken (1958) provided a further summary of results from the Gulf of St. Lawrence and southwestern Nova Scotia tagging investigations from 1946 to 1953. He noted that recaptures of halibut released in the LaHave-Roseway region supported the view that the stocks of LaHave-Roseway and Banquereau Bank intermingled but remained distant from other banks. Halibut recovered in the late summer fishery off Cape Sable Island originated from German and Browns Banks rather than from the eastward. He noted further that seasonal movements from deep to shallow water or the reverse were not apparent in this region from the tag recoveries. Halibut recaptured in relatively shallow water off Cape Sable Island originated at about the same depths on banks to the westward.

Wise and Jensen (1959) reported on the movements of tagged halibut off New England, based on recaptures from 1956 to 1958. In their study, 24 fish were tagged in the western part of the Gulf of Maine, 1 on Georges Bank, 15 near the entrance to the Bay of Fundy, and 151 on Browns Bank. Seventeen marked halibut were recaptured by the date of their paper and most were recovered close to the tagging areas. Four fish originating from Cashes Ledge were at large up to 55 weeks and were recaptured within 50 km of the tagging area. Two halibut, larger than the average, showed more extensive movement. One, tagged on Browns Bank in March 1957, was recaptured 8 weeks later and 160 km away on central Georges Bank. The other, the single marked fish released on Georges Bank (April, 1957), was recaptured 17 weeks later 96 km away on northern Browns Bank. Wise and Jensen felt that the fish had crossed the deep water of the Northeast Channel rather than following isobaths along the coast, as the latter would have entailed a migration of great distance. Tåning (1938) also indicated that migrations across deep channels occur.

Jensen and Wise (1961) provided an update of their tagging experiments based on two recoveries from an additional release of three and eight marked fish on Jeffrey's Ledge and Browns Bank, respectively, and a further 13 recaptures from their 1957 work. Two fish from the 1957 tagging were taken only a few kilometres from their site of release, despite having been at large for 80 and 109 weeks. In contrast, nine recaptures from tagging done on Browns Bank showed considerable movement to the northeast, with four recaptures from the Grand Bank, one from Banquereau Bank, and four from Sable Island area. Data appearing in the 1961 paper do not bear out the authors' earlier contention (Wise and Jensen 1959) that larger fish move greater distances, on average. However, the sample comprised relatively few fish. The authors could not conclude that halibut undertake substantial migrations and indicated that individual fish movement seems almost a matter of chance.

Kohler (1964) noted that recaptures from the 1962 tagging experiment in the Gully region of the eastern Scotian Shelf were mainly from the area of tagging during 1962. In 1963, however, recaptures showed that the predominant movement was to the east of the Laurentian Channel, to the area south of St. Pierre and Green Banks and along the southwestern edge of the Grand Bank. Of the remaining 15 recaptures, 3 were taken well east of the tagging area, 2 on the eastern edge of Banquereau, and 1 near Scatari Island.

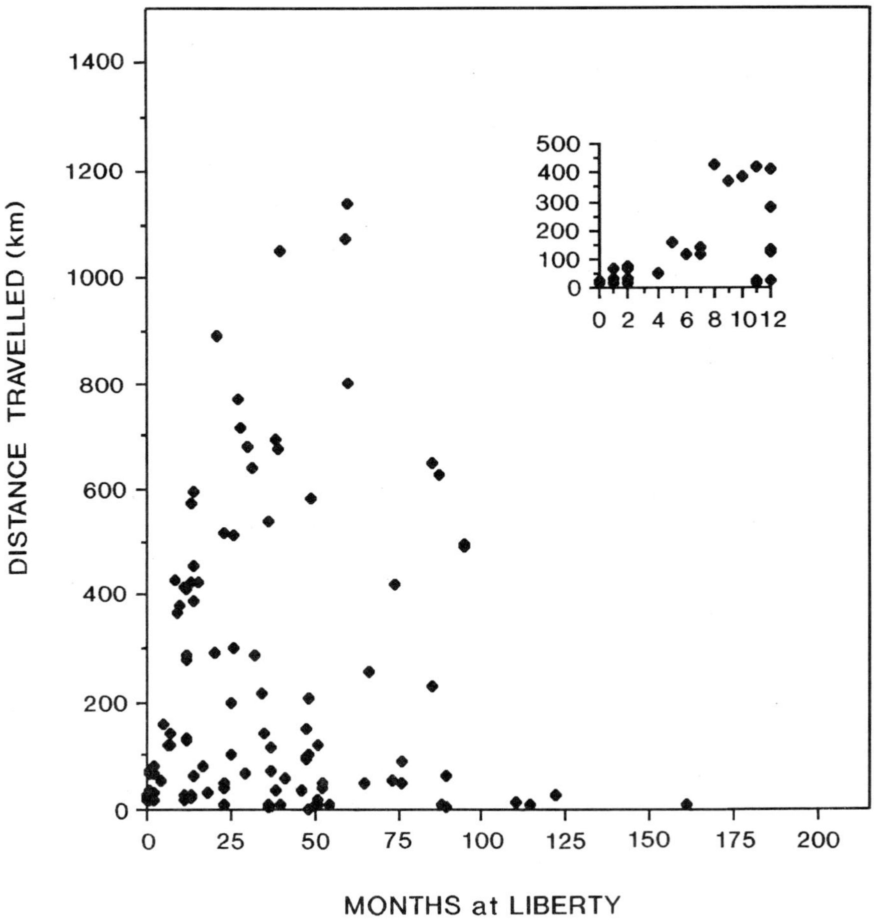

Fig. 15. Distance travelled versus months at liberty for Atlantic halibut released in the vicinity of Sable Island Gully, March 1962. The inset provides detail on distance travelled during the first 12 months after release (from Stobo et al. 1988).

More recently, Bowering (1986) noted that seven Atlantic halibut (all <58 cm in length) tagged off southwest Iceland were recaptured in Newfoundland waters (four fish) and off West Greenland (three fish). Of particular note is the comparatively small size of these fish, supporting the view given earlier that the small fish disperse to a greater extent than do the adults. Bowering concluded that while localized spawning may occur in both the Scotian Shelf and the St. Pierre - Grand Bank areas, the apparent high degree of mixing of adults evident in the studies by Jensen and Wise (1961) and Kohler (1964) may warrant the consideration of combining the areas for stock assessment purposes.

Stobo et al. (1988) also postulated that a return migration of mature fish occurred in the northwest Atlantic. They reached that conclusion from the observation that while fish tended to disperse from the tagging site with time, the majority were recaptured within 200 km of the release site, with some captured much closer (Fig. 15). Random dispersal, a hypothesis supported by Jensen and Wise (1961), could have explained such a result. However, Stobo et al. (1988) proposed an alternative hypothesis: the extent of movement of Atlantic halibut is negatively correlated with fish length and larger sexually mature fish exhibit limited dispersal and an annual return migration. They postulated that if the hypothesis is correct, many of the fish recaptured close to the Sable Island Gully release site may have come back to spawn, an interpretation seemingly supported by the pattern of returns of larger fish in Fig. 16, with fish captured 8-12 months later generally showing little or no apparent movement compared with returns made earlier. Further evidence came from a study of Kohler (1967), who found halibut in both the spent-recovering and ripening stages of gonad development near the area of tagging in February-March. Since the main tagging experiments occurred during that period, many of the adult halibut reported here would have presumably been in a spawning condition and near their spawning grounds. Kohler's findings, therefore, lend credence to their inference of a return migration to spawn.

Results such as those of Stobo et al. (1988) and Godø and Haug (1988) present somewhat of a paradox. If homing is a widespread phenomenon among halibut populations, a considerable amount of reproductive isolation would be a natural consequence. Such isolation is inconsistent with the results of the electrophoretic analyses reported by Grant et al. (1984), which suggest very little differences in the genome of Atlantic halibut populations, at least east of Greenland.

The homing by halibut to summer feeding grounds reported by Skud (1977) and to winter spawning grounds reported by Skud (1977) and Godø and Haug (1988) suggests that the adult fish establish stock (fishing) units that would be independent of fishing on other stock units. However, data and analyses are not yet available to determine if such separations extend to population (genetic) units, which is subject of considerable speculation.

Homing and attendant reproductive isolation would promote the development of separate populations. The unidirectional countermigration of juvenile halibut, at least in the Pacific, suggests a genetic programming. For example,

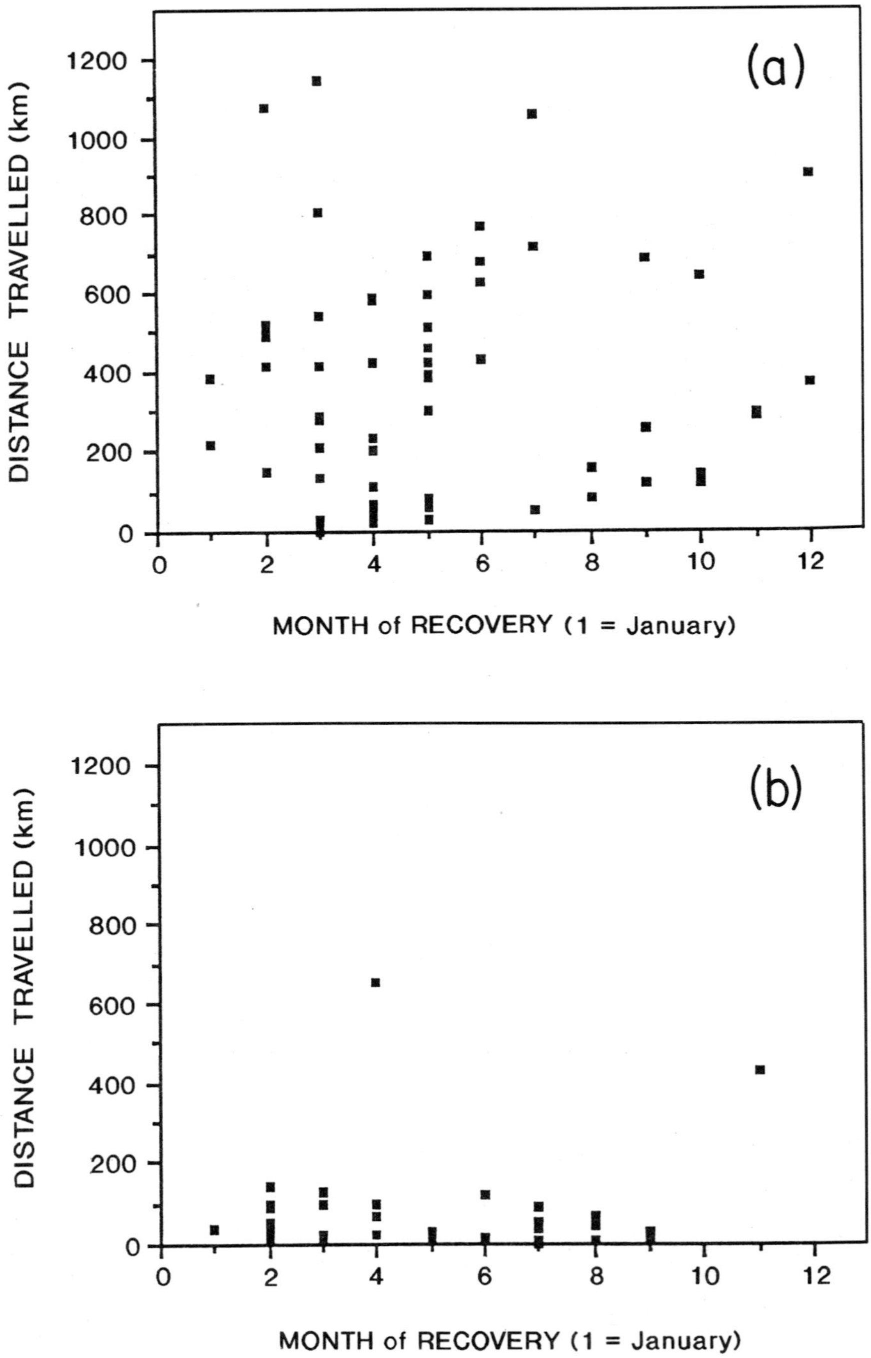

Fig. 16. Distance travelled versus month of recovery for Atlantic halibut released in the vicinity of Sable Island Gully, March 1962 (from Stobo et al. 1988) *(a)* Fish less than or equal to 75 cm. *(b)* Fish longer than 75 cm.

progeny of Pacific halibut spawned off northern British Columbia or southeastern Alaska would tend to drift with ocean currents in such a way that countermigration from nursery grounds would bring juveniles back to the vicinity of the spawning grounds. The greater the separation between spawning areas, the lower the probability of mixing of progeny.

The tendency to isolate would weaken as the range of drift and countermigration increases. Alternating strong and weak current regimes would cause metamorphosing larvae of different spawning areas to overlap on nursery grounds. Countermigration proportional to biotic and physical stimuli, such as directly with density of other juveniles or inversely with current velocity, would also increase mixing. Mixing does not need to be strong enough to cause uniformity, only enough to keep genetic input high enough to prevent isolation. Movement of juvenile Pacific halibut from the Bering Sea to the coast of Washington-Oregon-California shows potential for substantial mixing. Mixing during larval drift and juvenile dispersion may also account for genetic homogeneity observed for Atlantic halibut.

Age and Growth

Halibut are the largest of all flatfish and are among the larger species of fish in the sea. Both halibuts generally exhibit sexual dimorphism, and females grow substantially larger than males. The largest specimens in the Atlantic and Pacific are over 3 m long and have been reported to weigh 300 kg; these weights have not been thoroughly documented. Age, length, and weight of Pacific halibut are obtained from otoliths, collected from sampling the commercial fishery and from research cruises. Age is derived directly from counting annuli, and fish weight can be estimated by proportion to otolith weight or fish length (Quinn et al. 1985).

Otoliths of Atlantic and Pacific halibut have some marked dissimilarities, even when viewed macroscopically (Fig. 17). Harkonen (1986) notes several important characters of Atlantic halibut otoliths: the shallow sulcus is elevated above the surface of the proximal surface; the anterior-most point of the cauda (a small depression within the posterior end of the sulcus) is clearly anterior of the posterior-most point of the ostium, another marked depression, which occurs near the anterior end of the sulcus; and the ostium is connected to the edge of the otolith by a narrow groove. Our observations of the proximal surfaces of the Pacific halibut indicate similarities with the Atlantic form. For example, the ostium is also connected to the edge of the otolith by a shallow groove. The orientation of the ostium with respect to the cauda appears similar. However, otoliths of the Atlantic species tend to be larger with age, particularly after 8 years of age. The rostrum of the otolith of the Pacific form seems initially more pronounced, but in older fish the difference is reversed. The notch between the rostrum and the anti-rostrum appears more pronounced in the Pacific species. Such differences lend credence to the view that the Atlantic and Pacific forms are separate species, as the otolith shape is considered to be a species-specific characteristic (Frost and Lowry 1981). However, our inferences concerning stock discreteness must be viewed with caution, as the sample

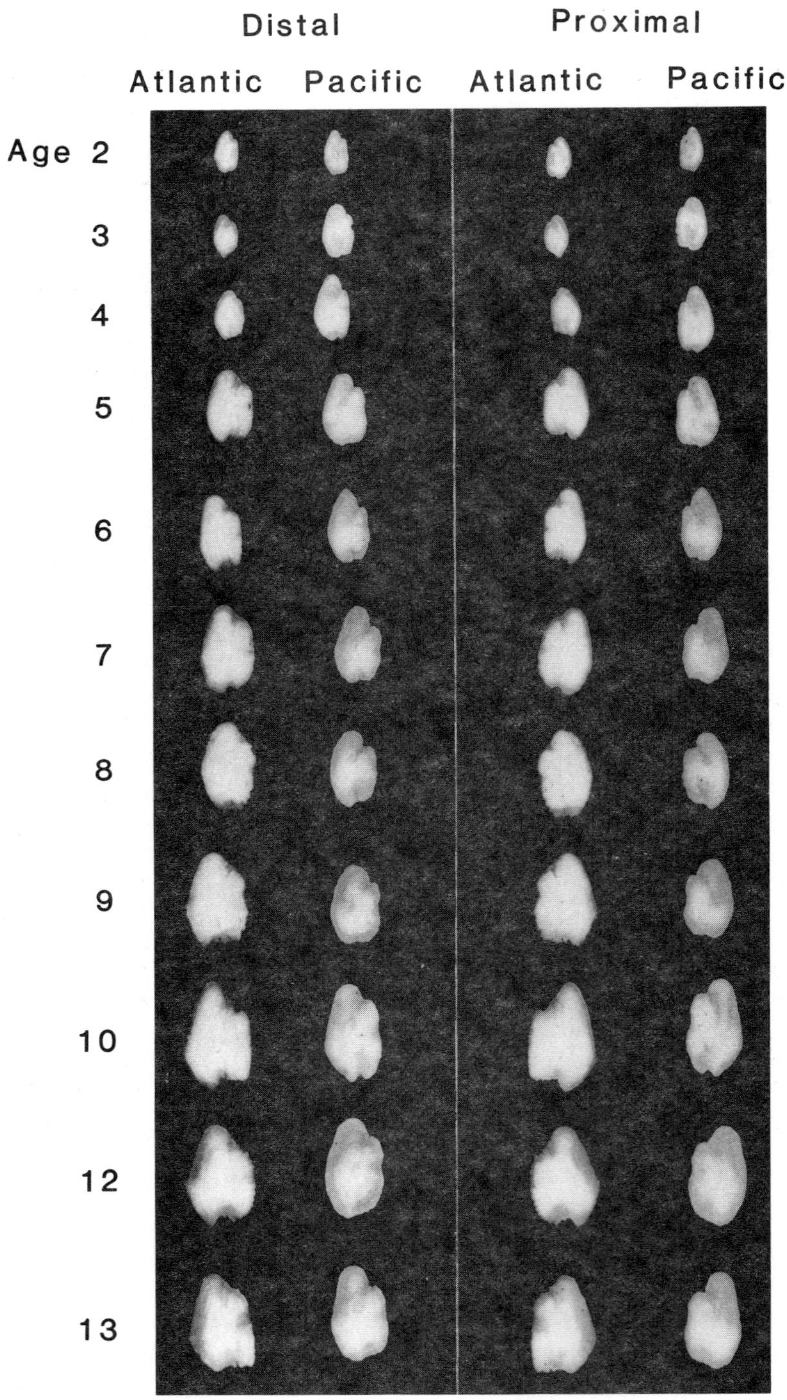

Fig. 17. Comparison of gross morphology of sagittal otoliths of Atlantic and Pacific halibut, with both proximal and distal faces of the otoliths shown.

size was small and in the case of Pacific halibut, considerable variation in otolith dimensions has been noted among samples collected from different geographic locales (C. L. Blood, IPHC, Seattle, Wash., personal communication). Confirmation of rings on halibut otoliths as true annuli is being tested for Pacific halibut through release of tagged fish injected with oxytetracycline (OTC). Fish absorb OTC during deposition of new bone, placing a time-mark on the otolith. In 1982 and 1983, 1791 injected fish were released and 1455 uninjected fish were released as a control. Through 1989, tags from 256 OTC-marked Pacific halibut have been recovered, 139 with associated otoliths. Comparison of the time at liberty to the number of rings laid down since release verifies the rings as annuli (C. L. Blood, IPHC, Seattle, Wash., personal communication). While age readers tended to slightly over age the otoliths, 95% of the readings were within 1 year of time at liberty. The IPHC uses surface reading, rather than breaking and burning the otolith, to estimate Pacific halibut age. Blood's study demonstrates that surface readings are accurate for the 8- to 15-year-old fish that dominate the commercial catch. The study did not evaluate accuracy of surface readings for rare, older fish. Age validation studies for northwest Atlantic halibut are under consideration.

The oldest age recorded for a Pacific halibut is 42 years for females and 27 years for males (IPHC 1987). A 50-year-old Atlantic halibut was reported from the continental shelf of the eastern Atlantic near the Faroe Islands (Jákupsstovu and Haug 1988). Most halibut in the Pacific setline fishery are 10-14 years old, while the Atlantic fishery harvest is predominantly in the 6- to 12-year-old range (Fig. 18). Age composition in the commercial fishery for the two taxa reflects the exploitation pattern of fishing and the selectivity of the gears used. Older ages result in the Pacific fishery because trawl retention is not permitted, longline gear selects for larger halibut than trawls, and a minimum size limit of 81.3 cm is in force. No size limit for Atlantic halibut was in effect at the time these data were collected for either trawl or longline, and overexploitation in the Atlantic may have reduced age composition.

Kask (1935) first analyzed tagging data to determine Pacific halibut growth for combined sexes. He noted that growth analysis was confounded by lack of sex composition of commercially caught fish,[2] and by comparison of dissimilar size ranges because females grow faster than males. Southward (1962, 1967) used back-calculated lengths from otoliths to demonstrate a change in growth rates of Pacific halibut during the 1950's and 1960's. Schmitt and Skud (1978) noted similar changes while examining fecundity at age.

McCaughran (1981) developed statistics for modelling Pacific halibut growth, using a subset of tagging data that included sex of recaptured fish. The natural logarithm of lengths converted the data to a normal distribution. The three-parameter von Bertalanffy equation and a two-parameter power function fit the data equally well when tested with tag return length data for males from the northern British Columbia

[2] No sex composition data are available from the Pacific halibut commercial fishery because the fish are eviscerated at sea and the gonads are discarded.

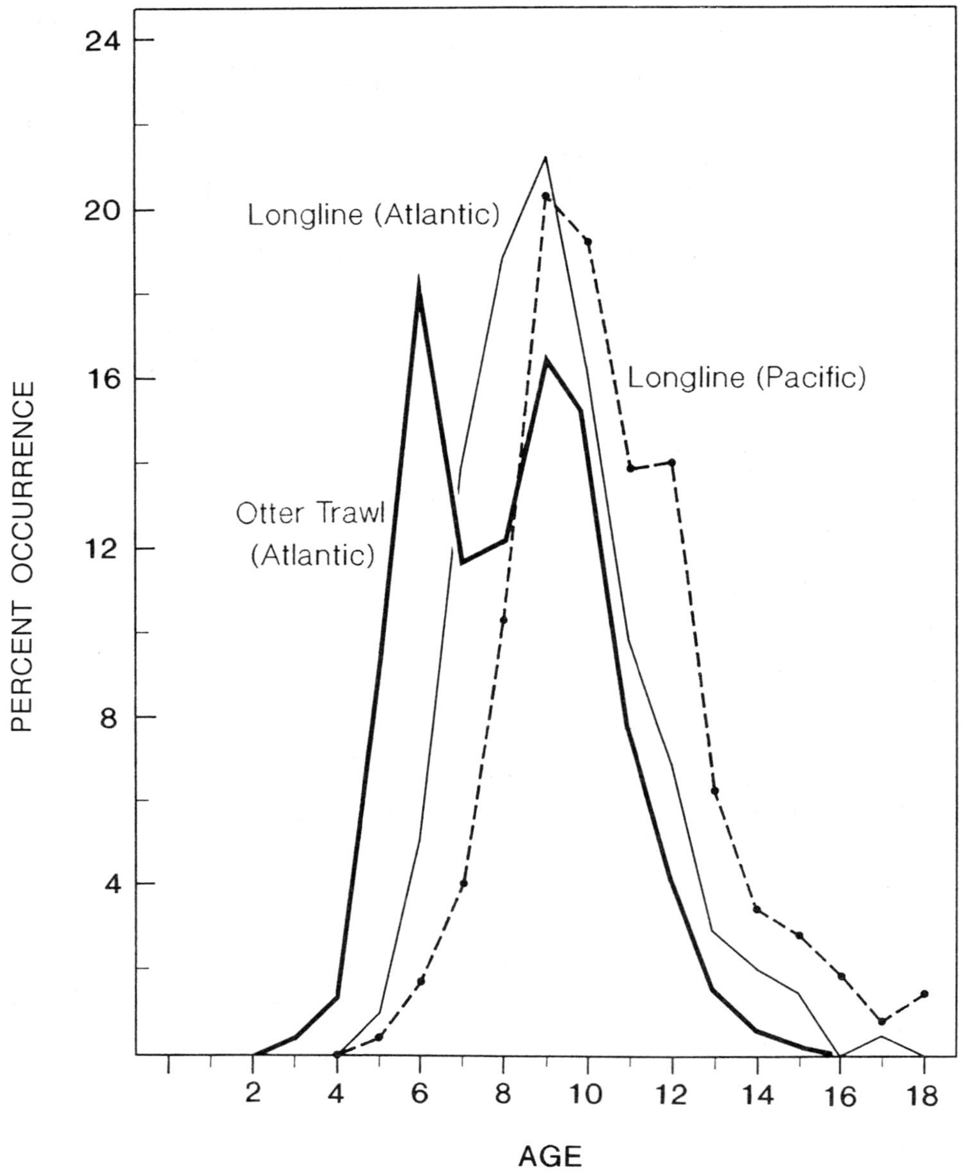

Fig. 18. Age composition of landings in the halibut fisheries. The Atlantic fishery data includes both longline and trawl samples for 1985–1987, whereas the Pacific data is from the longline fishery only for 1988.

coast. Growth parameters were subsequently estimated for females using the power function, which was preferred for its simplicity. McCaughran (1987) extended his previous analysis to males and females throughout the range for year-classes 1935-65. To avoid gear selectivity bias, longline-caught fish less than 10 years old when tagged were excluded from the analysis. Growth of either sex varied little since 1935.

The apparent discrepancy between the conclusion by Southward (1962, 1967) and Schmitt and Skud (1978) of changed growth rates over time and that by McCaughran (1981, 1987) of no change is attributed to length measurements used.

The former studies used estimated or measured lengths of individuals of a wide size range, while McCaughran used length increments determined from tagging of large individuals. The results of these authors are consistent, and indicate that change in growth rates occurs at younger ages, perhaps from density dependent or environmental influences. By the time Pacific halibut reach larger sizes, growth increments are fixed.

Mean length at each age is used to compare growth of Pacific and Atlantic halibut. This procedure only approximates true growth, because the samples were collected by longline and trawl fishing gears. These gears have select specific sizes, and consequently the data used here may not represent existing populations in the areas fished. In addition, fish from populations subjected to intensive fishing pressure over long periods by selective fishing gear may not represent growth rates in an unexploited state. For Pacific halibut, for example, longline gear selects fastest growing individuals that reach the legal size limit first; the remaining population contains slower growing individuals.

Acknowledging these problems, we compare data collected during the 1980's on average length at age for Pacific halibut from the western Gulf of Alaska (IPHC, Seattle, Wash., unpublished), for western Atlantic halibut from the Scotian Shelf and Newfoundland (J. D. Neilson, St. Andrews Biological Station, St. Andrews, N.B., unpublished), and eastern Atlantic halibut from the Faroe Islands (Jákupsstovu and Haug 1988).

Growth and sexual dimorphism of Atlantic and Pacific halibut are typically similar (Fig. 19). Growth is continuous, with little decrease at older ages. Faroese female Atlantic halibut grow to the largest size, while western Atlantic and Pacific females show little difference. Males from the Pacific, the eastern Atlantic, and the Scotian Shelf area of the western Atlantic diverge from the faster female growth pattern at 5-9 years of age. Faroese males appear slightly larger than Pacific males, and Scotian males are the smallest, with a marked growth decrease at older ages. Males from the Newfoundland area of the western Atlantic, however, are atypical of males in other areas, by exhibiting a growth pattern similar to that of the females (Fig. 19). Review of the data for Newfoundland halibut confirmed correct aging technique and sex discrimination, suggesting a fundamental, if unexplained, growth difference.

Trophic Interrelationships

No trophic information is available on larval life history stages. However, feeding by juvenile (post-metamorphosis) and adult stages of both halibuts has been investigated. In such fish, an often quite striking transition of prey selection from invertebrates to fish as halibut increase in size has been observed by many authors (Scott 1910; Jesperson 1926; McIntyre 1953; Rae 1958; Kohler 1967; Best and St-Pierre 1986). To compare the patterns of feeding, we reexamined data provided by Kohler (1967) and Best and St-Pierre (1986) for Atlantic and Pacific halibut, respectively (Fig. 20). While the data are not directly comparable (Kohler compiled simply "present/absence" data on each prey type, whereas Best and St-Pierre

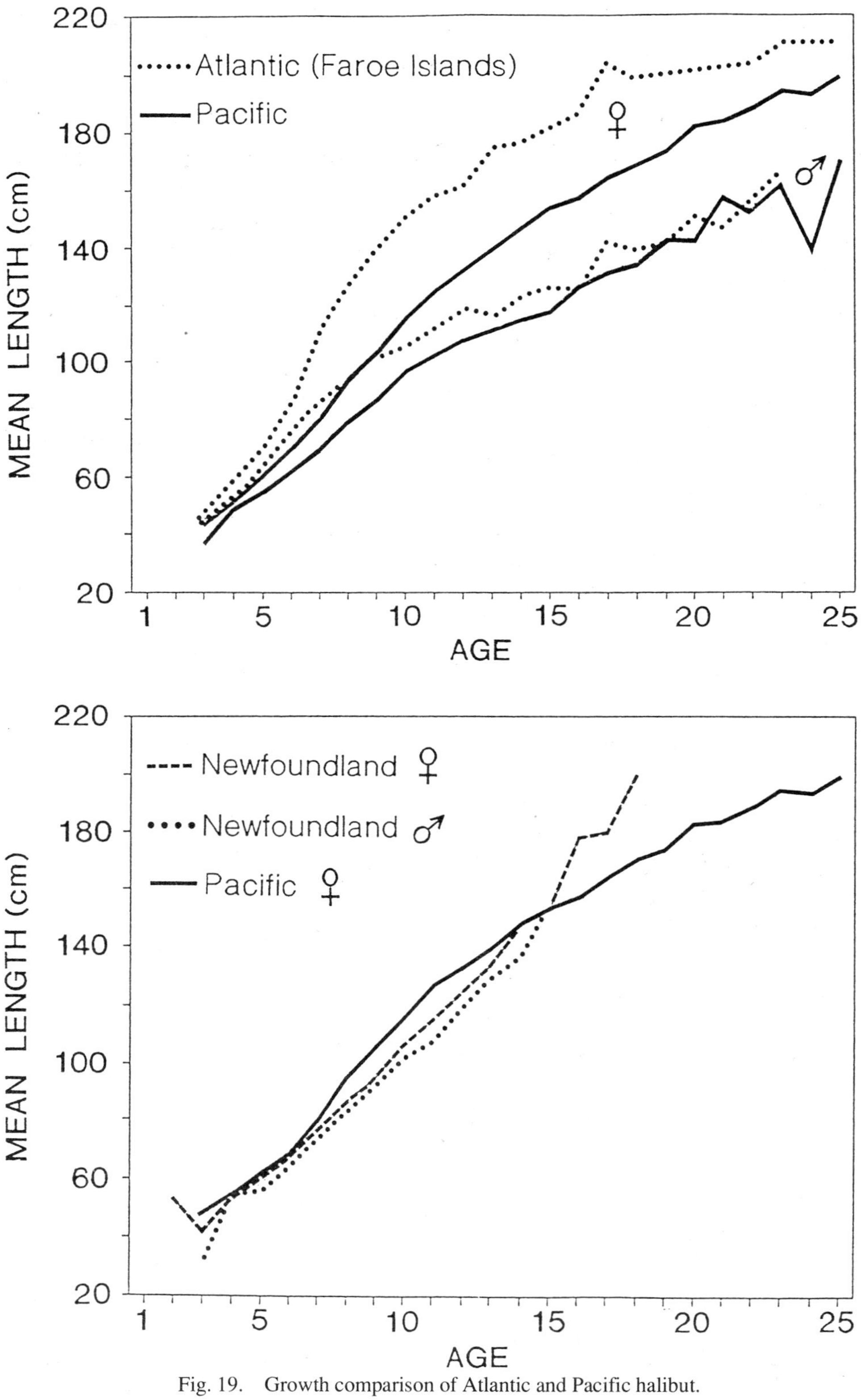

Fig. 19. Growth comparison of Atlantic and Pacific halibut.

enumerated all), a trend of increasing proportion of teleosts in the diet is evident for both Atlantic and Pacific forms. However, invertebrate prey appears to dominate the diet (by number) of Atlantic halibut more than for the Pacific fish.

Juvenile Atlantic halibut feed almost exclusively on invertebrates, while Pacific juveniles are strongly oriented to teleost prey. On the Atlantic side, small halibut (<30 cm) often feed on hermit crabs, prawns, small crabs, and mysids (Haug 1990). Rae (1958) also reports that among juvenile Atlantic halibut of lengths 21-50 cm, *Ammodytes* spp. were common prey; gadids (*G. aeglefinus* and *G. merlangus*) were also found, albeit less frequently than invertebrates. Unidentified shrimp and crab, Tanner crab (*Chionoecetes* spp.), and hermit crab (Paguridae) made up most of the invertebrate prey of Pacific halibut, while sand lance (*Ammodytes hexapterus*), walleye pollock (*Theragra chalcogramma*), and Pacific sandfish (*Trichodon trichodon*) constituted the most numerous teleost prey (Best and St-Pierre 1986).

As halibut grow beyond about 70 cm SL, the diet consists largely of other fish (Fig. 20). On the Atlantic coast of Canada, common prey include flatfish (including *Hippoglossoides platessoides*), redfish (*Sebastes* spp.), and pollock (*Pollachius virens*) (McIntyre 1952; Kohler 1967). Food items from 1 220 stomachs from the northeast Atlantic led McIntyre to conclude that halibut select prey. This conclusion was based on the observation that among the entire collection of halibut stomachs, only 33 species were found. Indeed, off Iceland, halibut were feeding on only 11 species, and fish (particularly *Sebastes marinus*) composed about 75% of the prey volume. In contrast, Homans and Needler (1944) found that haddock of the northwest Atlantic eat more than 200 species of prey. McIntyre further speculated that given the importance of *Sebastes* spp. in the diet of halibut, redfish could be considered the ecological analog of capelin and herring in the life history of cod.

McIntyre's (1952) conclusion that Atlantic halibut exhibit prey selection is at variance with some works on the feeding ecology of Pacific halibut. Best and St-Pierre (1986) concluded that halibut are opportunistic feeders, utilizing whatever is readily available: they examined over 2700 stomachs from juvenile halibut collected in the Gulf of Alaska and over 1300 stomachs from adult halibut from off British Columbia and Alaska and found that Pacific sand lance, walleye pollock, octopus, and Tanner crab made up a significant proportion of the halibut diet. A total of 30 and 40 different prey species were identified from the adult and juvenile collections, respectively. Brodeur and Livingston (1988) examined the gut contents of 301 Pacific halibut collected from two areas in the eastern Bering Sea. They concluded that fish supply about 80% of the food consumed on a weight basis, and that walleye pollock, yellowfin sole, and Pacific herring composed 48, 9, and 6% of the total prey volume, respectively. Using Odum's measure of prey diversity D (number of species consumed/$\log_{10}$ (total prey weight)), they concluded that halibut had a relatively diverse prey spectrum. Unfortunately, the authors did not stratify their results by fish size, reducing the value of their otherwise comprehensive analysis.

The apparent inconsistency of the conclusions on Atlantic and Pacific halibut prey selection is puzzling. If authors had been inconsistent in choosing the taxonomic level to which prey identification was completed, the difference in prey diversity

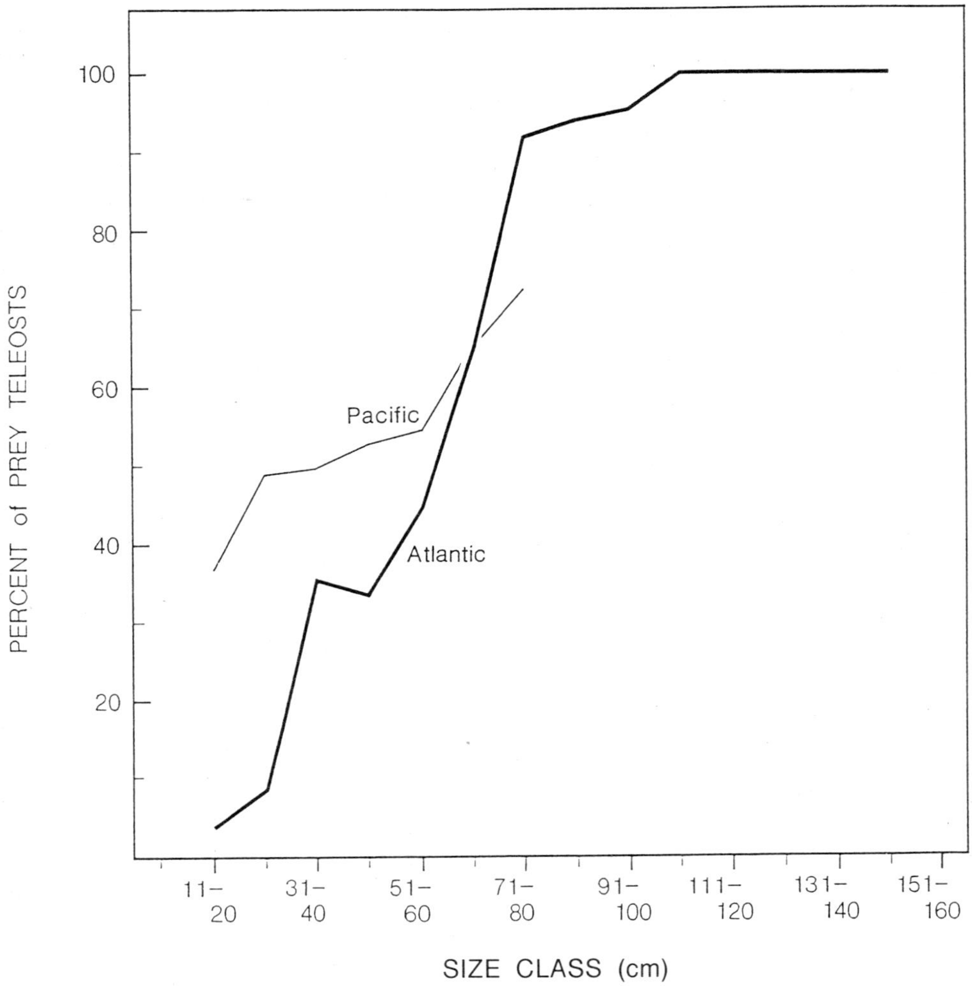

Fig. 20. Comparison of percentage of fish in the diet of Atlantic (Kohler 1967) and Pacific halibut (Best and St-Pierre) versus size.

might be considered an artifact; however, this explanation appears improbable, given that the identifications were carried to the species level whenever possible. The data of both McIntyre (1952) and Best and St-Pierre (1986), indicating that only a few fish species provide a large fraction of the food for adult halibut, supports the tentative view that for adults at least, some selectivity is present.

There is some evidence of cannibalism in both Atlantic and Pacific populations. In stomachs of 250 large (120 cm) Pacific halibut collected on nursery areas, about 7% contained remains of smaller conspecifics (Best and St-Pierre 1986). Halibut cannibalism on larvae was observed in 1986 at stations close to shore in the Gulf of Alaska. Eleven 3- and 4-year-old halibut examined for stomach contents contained 28 (0-8/stomach) stage 11 and 12 halibut postlarvae (IPHC 1987). On the Atlantic side, McIntyre (1952) reported that halibut composed 1% by volume of stomach contents from 33 halibut taken off Iceland. However, these specimens were small (less than 31 cm in length) and as has been shown earlier, such fish are not often piscivorous. Other reports from the Atlantic indicate either low or no incidence of cannibalism (Scott 1910; Joensen 1954).

Halibut are only rarely found as prey of other fish (Best and St-Pierre 1986). Marine mammals, notably the Stellar sea lion (*Eumetopias jubata*), are notorious for preying upon Pacific halibut hooked on setline fishing gear, often forcing fishermen to change their grounds (Best and St-Pierre 1986). Those authors also provide a review of the incidence of feeding by other marine mammals on Pacific halibut and concluded that the rates of mortality were low. They concluded that this was due to the size and active nature of Pacific halibut, along with their bottom dwelling habits, which renders them less accessible to predation than schooling pelagic species. Similarly, a study of 279 harbor seal stomachs by Boulva and McLaren (1979) revealed no evidence of predation on Atlantic halibut.

Parasites

In general, the parasitic fauna of Atlantic and Pacific halibut appear incompletely described. Scott and Bray (1989) studied the helminth parasites of the alimentary tract of Atlantic halibut and concluded that both prevalence (percentage of fish infected) and intensity (mean number of parasites per infected fish) were low. A total of 23 helminth parasites were identified (16 Digenea, 3 Cestoda, 2 Acanthocephala, and 2 Nematoda). The authors found a number of new host records, reflecting the lack of this type of research on Atlantic halibut. They further concluded that changes in parasite prevalence and intensity with latitude were in the form of clines and were consistent with the interpretation of stock structure presented by Stobo et al. (1988).

A project investigating the protozoan and metazoan parasites of Pacific halibut, undertaken in 1990 (R. B. Blaylock, Pacific Biological Station, Nanaimo, B.C., personal communication), found 29 parasite species. The number of species per fish ranged from 1 to 8 (mean = 4) and the number of worms per fish ranged from 2 to 1242 (mean = 176). The majority of alimentary tract parasites were Trematoda. Among other parasites found, *Anisakis simplex* and *Pseudoterranova decipiens* are capable of infecting humans if improperly cooked fish is consumed. Presence of the muscle-dwelling protozoan *Kudoa thrysites* is related to "milky flesh" if halibut are not frozen soon after landing. Significant differences among localities occurred for number of worms, number of species, and diversity; the Bering Sea accounted for most of the differences. Analysis suggests that halibut parasite communities may have a baseline predictability. Some species appear to be more common in halibut than in other marine fish.

History of the Fisheries

Native North Americans fished for halibut in the Atlantic and Pacific Oceans before the arrival of European explorers. Little anthropological information is available for the Atlantic. Research at archaeological sites in northern Maine show very few bones of Atlantic halibut, even though other fish remains number in the thousands (David Sanger, University of Maine, personal communication). Several vertebrae drilled for jewelry accounted for most of the bones. Sanger concluded that either Atlantic halibut was not a prime species from the Indian perspective or they were too difficult to harvest with existing technology.

The methods used by Pacific northwest Indian tribes to fish for Pacific halibut are well documented in the journals of early explorers and settlers during the late 18th through mid-19th centuries. Summaries of the early reports prepared during the 1970's and 1980's provide access to the material.[3] Specialized lines, floats, and hooks developed by the Indians (Fig. 21) were very effective (Stewart 1977), and out fished gear used by early explorers from Europe. Members of coastal tribes used halibut for subsistence and bartered halibut to tribes without access to the fish (Bell 1981). The Indians continued bartering halibut to early settlers. From 1855 to 1860, several Indian tribes in Washington state signed treaties with the U. S. government that, among other things, reserved fishing rights for the tribes. The U.S. government has since upheld treaty rights to fish for several species, including Pacific halibut.

Descriptions of the early non-Indian halibut fishery off the northeastern United States and Canada (Goode and Collins 1887; Scudder 1887; Bigelow and Schroeder 1953) give the impression that halibut were of low value and a nuisance to the Atlantic cod fishery throughout the early 1800's. Significant commercial halibut fishing began on the U.S. Atlantic coast during the 1820's, in the Massachusetts Bay - Cape Cod area in response to market demand (Zwanenburg 1984). By the 1830's, with catch rates declining in these near shore areas, the fishery expanded to the offshore banks. Halibut were generally considered very abundant throughout waters off northern New England and the Canadian Maritimes. As the U.S. market for halibut improved around the mid-1800's, the halibut fishermen originating in the New England area travelled as far afield as the Grand Banks of Newfoundland by 1864-1866; to the west Greenland banks by 1866, to Magdalens by 1873, to the northern part of the Gulf of St. Lawrence by 1881, and to Iceland by 1884 (Bigelow and Schroeder 1953).

During this period and through the late 1800's, Atlantic halibut was not a preferred food fish in the Canadian fishing industry and any bycatch was either

[3] For example, the Makah tribe of northwest Washington state was one of the foremost halibut fishing tribes on the Pacific coast. Their halibut fishing history was summarized in Lane (1977). Makah Marine Navigation and Traditional Offshore Fisheries. The report was prepared for the Makah Tribe and submitted to the U.S. Federal District Court, Tacoma, Wash.

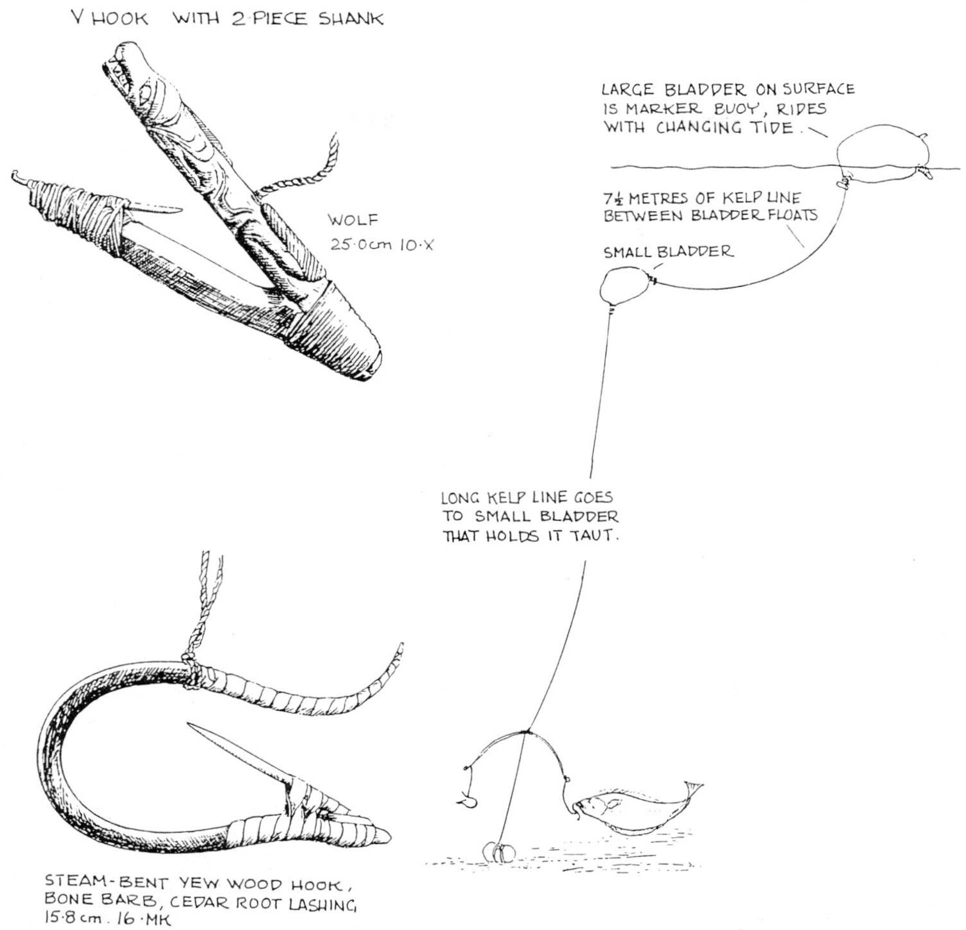

Fig. 21. Examples of northwest coastal Indian halibut fishing gear (from Stewart 1977, with permission).

discarded or used as bait. At this time, Pacific halibut was used locally but commercial fishing was limited.

By the late 1800's, both Canadian and American fishermen were directing their efforts toward Atlantic halibut on the Scotian Shelf, Newfoundland Grand Banks, and the Gulf of St. Lawrence (McKenzie 1946; Zwanenburg 1984). American vessels also carried out a lucrative Atlantic halibut fishery as far north as the Davis Strait - West Greenland area until the late 1890's when it became unprofitable (Devold 1938). Figures 22*a*-22*c* shows typical scenes of the early days of the Atlantic halibut fishery.

The non-Indian commercial harves of Pacific halibut began off the coast of Washington in 1888 with a landing at Tacoma of 30 t[4] which was sent by rail to the

[4] Pacific halibut landings have been traditionally measured in millions of pounds net weight (headed and gutted, ice and slime removed). The standard conversion used by IPHC from round weight to net weight is 0.75.

Fig. 22. Scenes from early days of fishing for Atlantic halibut off North America. Courtesy of the Maritime Museum of the Atlantic, Wallace Collection, Halifax, N.S. (*a*) Dorymen from the schooner F/V *Albert J. Lutz* clubbing Atlantic halibut. (*b*) Atlantic halibut of about 135 kg from the Anticosti area aboard the schooner *Albert J. Lutz*. (*c*) Unloading Atlantic halibut at Lunenburg, N.S. from the schooner F/V *Leah and Shirley*.

east coast. Fishing areas expanded from the waters off Washington and Oregon, as apparent depletion of the grounds occurred, to British Columbia and ultimately to Alaska. Figures 23*a*–23*d* show typical scenes of the early days of the Pacific halibut fishery.

In the early 1900's, some directed Atlantic halibut fishing by Canadian and American vessels occurred in some of the deep channels off Labrador and the northeast Newfoundland Shelf; however, these operations did not continue into the second half of the century as stocks declined (Bowering 1986). The only areas with sufficient abundance to allow a significant contemporary Canadian fishery for Atlantic halibut are along the southern edges of the Newfoundland Grand Bank, the southwest slope of St. Pierre Bank, the channels between St. Pierre Bank and the Newfoundland Grand Bank, the Scotian Shelf, and to a lesser extent, the Gulf of St. Lawrence. There are, therefore, no longer any Canadian directed fisheries for Atlantic halibut north of the southern Newfoundland Grand Banks, nor American fisheries in the southern range of the distribution. In contrast, Pacific halibut is fished throughout its range in spite of fluctuating abundance.

When the commercial fishery began, Atlantic and Pacific halibut were caught using baited handlines in relatively shallow water (less than 140 m) fished from small boats such as dories. In the Pacific, dories were deployed from sailing vessels, three of which began the commercial fishery in 1888, using two-man dories and hook-and-longline gear for actual fishing. As the fisheries developed and expanded to offshore and more distant areas, handlining was replaced by longlining and this method has remained much the same as it began in the mid-1800's. Pacific halibut longline gear consists of a ground line, usually 0.54 km long, with hooks attached to gangions spaced 3-10 m apart along the ground line (IPHC 1987). Traditionally, gangions have been tied to the ground line at a set spacing ("conventional" gear), but more recently gangions may be attached to the ground line with a snap fastener ("snap" gear). Conventional gear is set and retrieved as coils, while snap gear is set and retrieved on drum. Several ground line units, called skates, are strung together for a fishing unit. Myhre and Quinn (1984) found no intrinsic difference in the efficiency of the two gears, although newer entrants into the fishery and smaller vessels tend to use snap gear, which had a lower catch per unit effort than vessels using conventional gear.

In the Pacific, sailing vessels and dories gave way by the 1920's to gasoline- and diesel-engine powered vessels, which hauled longlines directly from the deck. While hydraulics and navigation improvements have made longline fishing more efficient on both coasts, one of the biggest changes was the replacement during the mid-1980's of the traditional J type hook with more efficient circle hooks (Fig. 24), which have catch rates more than twice the rates of the J hook (IPHC 1985).

On the Pacific coast through the 1960's, the fleet comprised about 200-300 longline vessels fishing almost exclusively for Pacific halibut. Limited entry in other fisheries, collapse of king crab stocks, and increased profitability (especially since 1980) in the Pacific halibut fishery caused an increase to 4264 active vessels and an additional 2270 registered but inactive vessels in 1990 in the U.S. In 1990, 435 vessels

Fig. 23. Scenes from early days of fishing for Pacific halibut off North America. Courtesy of the International Pacific Halibut Commisssion, Seattle, WA. *(a)* West coast Indians cleaning Pacific halibut on the beach about 1910, probably near Cape Flattery, Washington. Coastal Indians fished from canoes. *(b)*Baiting up on the steamer F/V *Grant* during the era of dory fishing (dories nested in the background). The *Grant* was built around 1905 and fished for Pacific halibut until sunk off the Canadian coast in 1911.

(c) Dressing a load of Pacific halibut on board the *Grant* circa 1910. The *Grant* had a hold capacity of about 110 t round weight (180 000 pounds net weight) of iced halibut. *(d)* The schooner F/V *Washington* tied up on the Seattle waterfront. The schooner fleet emerged during the 1920's. Noted for seaworthiness and endurance, many schooners are still active.

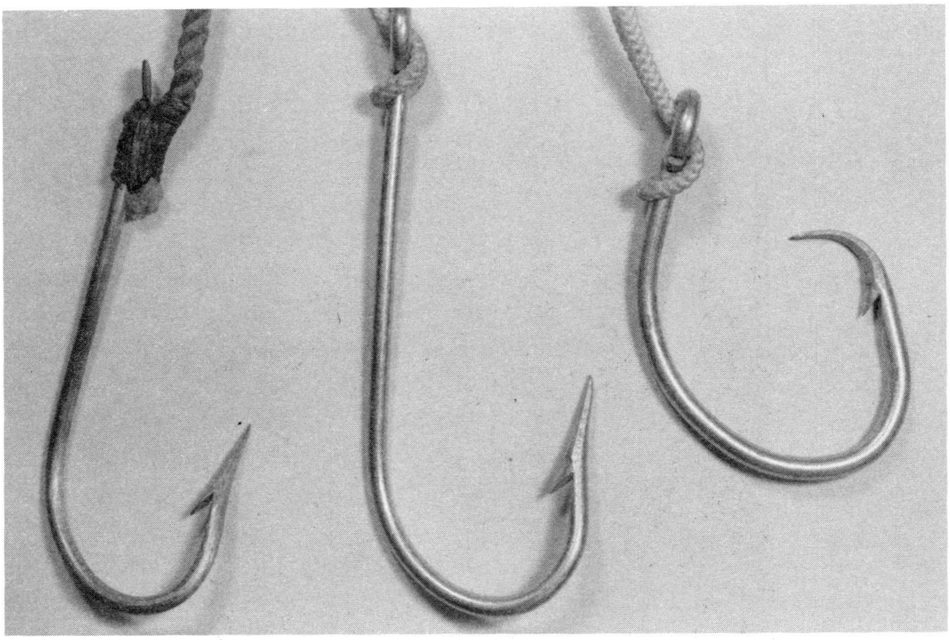

Fig. 24. Succession of hooks used in the halibut fisheries: left, early flattened J hook seized to the gangion with ganging twine; centre, eyed J hook that replaced the ganged hook in the 1960's; right, circle hook introduced in 1982–1983 (from IPHC 1987).

were eligible under Canadian limited entry to fish for Pacific halibut, and all obtained licenses.

While the Pacific halibut fishery has maintained a hook and line gear, other gears are used in the Atlantic fishery. Otter trawling began in the Northwest Atlantic around 1895 (von Brandt 1972), so halibut have probably been harvested to some degree since then. Until 1988, most western Atlantic halibut were caught by otter trawlers as a bycatch of other groundfish fisheries such as cod and haddock. Otter trawlers tend to catch much smaller halibut than those caught in the longline fisheries, which has prompted serious concern from the longlining component of the fishing industry. Trawl retention of Pacific halibut was prohibited in 1944, and regulations restricted retention of trawled Atlantic halibut off Canada to a small proportion of landed catch since 1988.

In recognition of treaty Indian fishing rights to Pacific halibut, the IPHC, in consultation with the U.S. Government, has authorized specific seasons or quotas to certain Washington tribes since 1985 (Trumble et al. 1991). Allocation to an expanded number of tribes currently occurs through the Pacific Fishery Management Council. Tribal harvest in Washington increased from about 2 t in 1985 to 92 t in 1989 before declining slightly in 1990. In 1990, the U.S. Government authorized a halibut fishery for the Metlakatla Indians of southeast Alaska, who harvested about 20 t. Tribal harvest is not monitored in other areas of the coast.

Fishing both halibuts for personal use has long been a common practice, but recreational fishing became important only during the 1980's. The recreational fishery is not managed nor is catch estimated for the Atlantic fishery. Recreational regulations were established by the IPHC in 1973 for Pacific halibut; until then, no fishing was allowed except during commercial seasons. The harvest was about 5% of the total directed harvest in 1990.

Fishermen from the U.S. and Canada historically fished off the coasts of either country beyond the 3-mile territorial limit. U.S. fishermen were prohibited from fishing off Canada in 1979, while Canadian fishermen were phased out of fishing in the Pacific Ocean waters of the U.S. in 1979, 1980, and 1981 under conditions of the 1979 Protocol to the 1953 Pacific Halibut Treaty, as required by extended jurisdictions of the U.S. and Canada in 1976. The two countries have a reciprocal landing agreement that permits fishermen of one nation to land Pacific halibut at ports of the other country. No such agreement exists for Atlantic halibut.

Population Dynamics

Trends in Catches

Some form of catch record history is available for both Atlantic and Pacific halibut since the late 1800's or early 1900's. However, reliable records extend for a shorter period. Thompson and Bell (1934) presented commercial catch data for Pacific halibut since 1888, but data prior to 1929 are not considered reliably reported or recorded (Myhre et al. 1977). Atlantic commercial data are considered most reliable since 1963, when Greenland halibut catch records were consistently separated from Atlantic halibut. Large fluctuations in commercial harvests occurred on both coasts. Since about 1910, Atlantic halibut landings ranged from approximately 1000 to 5000 t, while Pacific halibut harvest ranged from 12 000 to 45 000 t.

Catch records available since 1910 indicated an average annual catch of about 2000 t by Canadian fishermen up to the 1940's (Zwanenburg 1984). As a result of decreased effort during World War II, stocks rebuilt and yielded higher than average catches during the 1950's. Catch levels since then, including recent years, have been near the long-term average of 2000 t. Most accurate records of Atlantic halibut landings occurred from 1963 (Fig. 25), as earlier data would have also included Greenland halibut reported together with Atlantic halibut. This would not be a problem in the southern areas but could bias catch records in areas north of the Newfoundland Grand Bank and, to a lesser extent, the Gulf of St. Lawrence where Greenland halibut is more abundant. Only commercial catch data are maintained for Atlantic halibut.

The area of highest exploitation in the Northwest Atlantic is the Scotian Shelf area (Neilson et al. 1987) (Table 2). Catches here are almost entirely domestic, with an average of 79% of the catch since 1972 taken by longline and the remainder taken by otter trawls. For the St. Pierre Bank area, only 37% of the catch has been taken by longline since 1972, with the remainder taken by otter trawls. On the southern Newfoundland Grand Bank since 1972, 44% of the landings were attributed to longlines (W. R. Bowering, unpublished). Catches in the Labrador - eastern Newfoundland area are the result of incidental bycatch in other groundfish fisheries, mainly by otter trawlers. Catches in the Gulf of St. Lawrence also rarely exceed 200

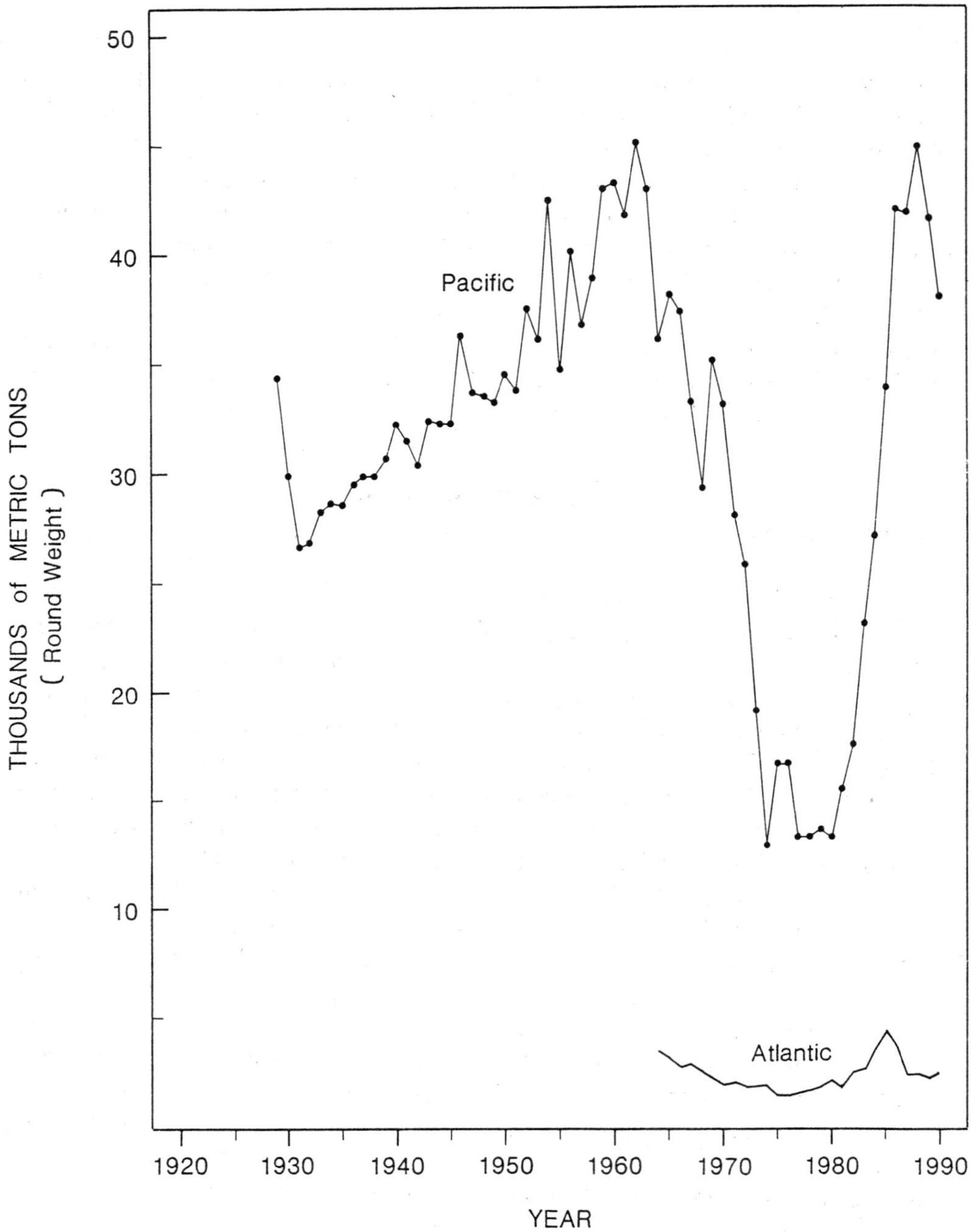

Fig. 25. Landings of Atlantic and Pacific halibut (t, round weight). The Atlantic time series is shortened because data are compromised prior to 1963 when Greenland halibut (turbot) were generically recorded with Atlantic halibut as "halibut."

t annually, although there is some directed effort by fishermen using longlines particularly in more recent years.

The commercial harvest of Pacific halibut has averaged more than 30 000 t per year, about 15 times the Atlantic halibut average. Annual fluctuations have been great over the course of the Pacific halibut fishery (Fig. 25). Commercial harvest

Table 2. Commercial Atlantic halibut catch by North Atlantic Fisheries Organization management areas (thousands of metric tons)

Year	Combined	4V	4W	4X	3Ps	3NO	2GHL 3KL	4RST
1976	1.417	0.144	0.148	0.422	0.087	0.258	0.162	0.196
1977	1.613	0.088	0.177	0.448	0.389	0.214	0.147	0.150
1978	1.635	0.244	0.283	0.565	0.165	0.180	0.063	0.135
1979	1.877	0.230	0.358	0.636	0.161	0.262	0.098	0.132
1980	2.113	0.339	0.371	0.748	0.145	0.150	0.158	0.202
1981	1.847	0.250	0.379	0.766	0.138	0.106	0.113	0.095
1982	2.488	0.342	0.476	0.907	0.171	0.351	0.150	0.091
1983	2.568	0.419	0.546	0.864	0.156	0.290	0.119	0.174
1984	3.378	0.496	0.572	0.924	0.311	0.742	0.157	0.176
1985	4.240	0.606	0.620	0.741	0.416	1.547	0.147	0.163
1986	3.877	0.553	0.579	0.500	0.472	1.298	0.202	0.273
1987	3.155	0.345	0.354	0.484	0.183	1.206	0.332	0.251
1988	2.824	0.298	0.360	0.608	0.142	0.888	0.273	0.255
1989	2.205	0.290	0.411	0.437	0.211	0.491	0.112	0.253
1990	2.367	0.274	0.365	0.446	0.218	0.502	0.059	0.503

gradually spread northward from the Washington-Oregon area to British Columbia and then to Alaska from the late 1800's to early 1900's. The majority of landings currently occur from Alaska waters, with the Gulf of Alaska the most important area (Table 3). British Columbia, the Bering Sea - Aleutian Islands, and Washington-Oregon-California areas follow with decreasing catches. The modern record high landing occurred in 1988.

Table 3. Commercial catch of Pacific halibut by International Pacific Halibut Commission management areas (thousands of metric tons)

Year	Combined	2A	2B	2C	3A	3B	4
1976	16.653	0.144	4.405	3.343	6.679	1.699	0.383
1977	13.226	0.125	3.282	1.927	5.226	2.010	0.656
1978	13.298	0.059	2.786	2.610	6.226	0.803	0.814
1979	13.624	0.028	2.937	2.740	6.855	0.236	0.828
1980	13.224	0.013	3.411	1.958	7.237	0.168	0.431
1981	15.565	0.122	3.170	2.719	8.587	0.252	0.717
1982	17.544	0.175	3.349	2.117	8.164	3.487	0.252
1983	23.214	0.220	3.283	3.998	9.423	5.096	1.196
1984	27.198	0.260	5.476	3.542	12.072	4.563	1.283
1985	33.937	0.298	6.280	5.571	12.611	7.538	1.638
1986	42.079	0.332	6.803	6.448	19.800	6.771	1.925
1987	41.902	0.358	7.389	6.483	18.789	6.183	2.700
1988	44.616	0.294	7.741	6.919	22.718	4.176	2.768
1989	40.509	0.284	6.326	5.764	20.400	4.742	2.982
1990	37.012	0.193	5.141	5.927	17.539	4.899	3.314

Pacific halibut may be retained only by longline gear during specified open periods; unlike the situation in the Atlantic, bycatch of Pacific halibut is not included in commercial catch records. Losses to the Pacific halibut resource from bycatch are estimated from bycatch rates and bycatch discard mortality rates in fisheries for other

Table 4. Components of total fishery removals for Pacific halibut (thousands of metric tons)

Year	Commercial catch	Bycatch mortality	Sport catch	Waste	Total
1976	16.632	8.028	0.181	0.000	24.841
1977	13.245	6.835	0.181	0.000	20.261
1978	13.306	7.117	0.242	0.000	20.665
1979	13.608	8.955	0.363	0.000	22.926
1980	13.245	11.024	0.484	0.000	24.753
1981	15.543	8.699	0.665	0.000	24.907
1982	17.539	7.196	0.786	0.000	25.521
1983	23.224	6.295	1.028	0.000	30.547
1984	27.216	5.875	1.149	0.484	34.240
1985	33.929	4.370	1.572	0.968	40.839
1986	42.094	5.011	2.117	1.935	51.157
1987	41.912	6.534	2.480	2.532	53.458
1988	44.634	8.580	2.480	2.113	57.807
1989	40.509	7.966	3.502	2.042	54.019
1990	37.012	10.547	3.608	1.986	53.153

species. Other sources of fishery-related exploitation of Pacific halibut include deadloss (waste) in the setline fisheries caused by lost or abandoned gear, or from mortality of undersized halibut returned to the sea, and catch by recreational fishermen (Table 4).

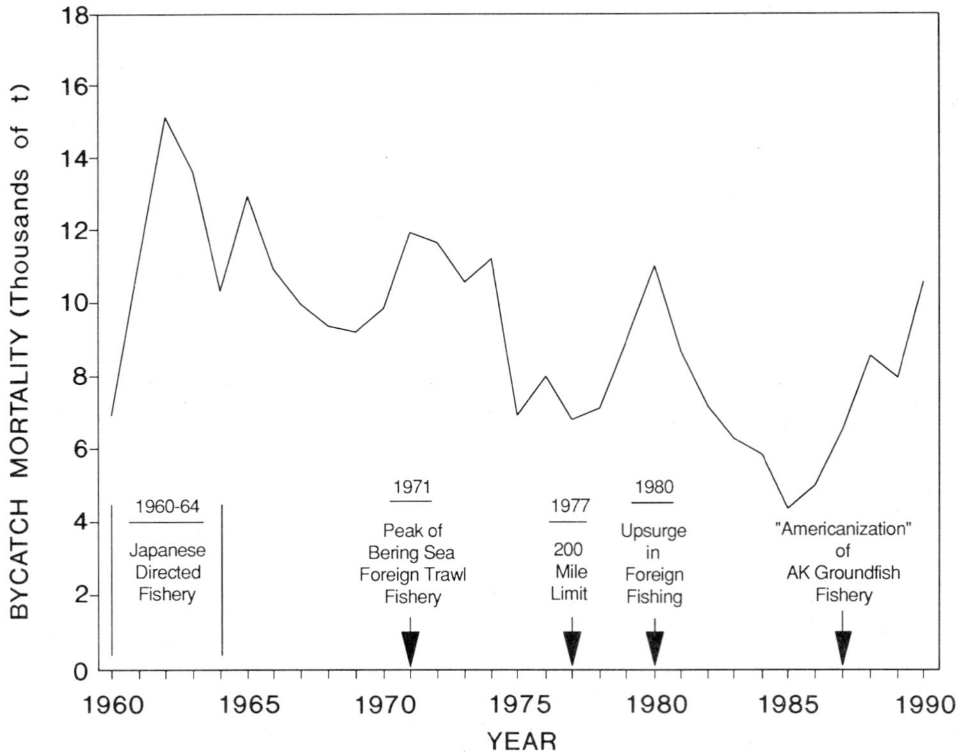

Fig. 26. Historical trend in estimated mortality of Pacific halibut caught incidentally in fisheries for other species or in foreign fisheries directed on Pacific halibut (from IPHC 1987).

Bycatch of halibut (and other prohibited species) is one of the major unresolved management issues in the North Pacific. Incidental mortality of Pacific halibut was relatively small until the 1960's, when it increased rapidly because of the sudden influx of foreign fishing vessels off the North American coast (Fig. 26). A short period of directed Pacific halibut harvest by Japan in the Bering Sea is included. The total incidental mortality peaked in 1962 at about 15 000 t. Halibut bycatch declined during the 1960's, but increased to about 12 000 t in the early 1970's. The bycatch dropped to a 8000 t level during the late 1970's and early 1980's. By 1986, the incidental mortality declined to 4000 t, the lowest level in recent history. However, incidental mortality has increased since 1986, reaching 10 000 t in 1990.

During recent periods of intensive fishing with short fishing periods in the Gulf of Alaska, Pacific halibut fishermen have maximized their harvest potential by setting more gear than can be retrieved during a fishing period, and often lose gear through tangles with the gear of other fishermen. Weather also causes gear loss. The IPHC has collected data on gear loss since 1986 and has used average catch per unit effort for the regions of loss to estimate wastage ranging from 1200 to 1900 t through 1990. Because wastage was also assumed to have occurred earlier, estimates for 1984 and 1985 were made as half the preceding year.

Sport fishing and subsistence use of Pacific halibut has occurred for many years (officially only since recreational regulations in 1973) at levels less than 300 t prior to 1979. The fishery is most active in the Gulf of Alaska, which accounts for about 75% of the sport catch. The sport harvest increased to approximately 3600 t in 1990.

Data Availability

The IPHC began to collect fishery data for Pacific halibut from the time of its inception in 1923, but statistics are most reliable since 1929 (Myhre et al. 1977). The Pacific coast of the United States and Canada was originally divided into 35 (now extended to 44) 96-km statistical areas which formed management areas when aggregated. Logbook information from fishermen provided distribution of catch, effort, fishing location, and auxiliary data by statistical area. Pacific halibut catch data recorded on landing documents (fish tickets) provided total catch. Age composition was derived from otolith samples collected from vessels that provided logbook data. While some of the earliest data are not suitable for detailed analysis, this data set is one of the longest and most complete for any fishery.

Comparatively few data are available on northwest Atlantic halibut population abundance. This situation is largely a result of the sporadic nature of the landings and that the fish are often landed with heads removed. Collection of data from observers stationed on board commercial fishery vessels has also been intermittent and incomplete in coverage, as most of the landings are made by longliners, and observers are most often deployed on trawlers. There is, however, a catch per unit effort series for longliners 20 m and larger since the late 1960's.

Initial Stock Assessment for Pacific Halibut

From the beginning of the IPHC, catch per unit effort (CPUE) has been a primary index of Pacific halibut abundance (Babcock et al. 1931). The unit of gear is the skate, which consists of groundline to which short lines with hooks are attached. At first, unadjusted changes in CPUE were assumed to directly reflect changes in abundance. Catch limits for the various management areas increased or decreased proportionately with CPUE. Skud (1975) reanalyzed the fishery data and showed that catch per skate did not accurately portray the biomass because catchability of the gear changed. CPUE remained the primary index of stock abundance through the early 1980's, but IPHC standardized CPUE to take into account such factors as fishing area, hook spacing, and hook type.

A measure of stock condition independent of CPUE seemed needed because not all factors affecting CPUE were known (Hoag and McNaughton 1978). Hoag and McNaughton applied Pope's (1972) method of cohort analysis to estimate stock size in numbers and fishing mortality. Biomass estimates were later obtained from CPUE estimates using the delay-difference model (Deriso 1980). Starting in 1981, biomass estimates from both cohort analysis and delay-difference models were used to estimate Annual Surplus Production (ASP), or the excess biomass production above that required to keep the population abundance at a constant level each year.

Current Stock Assessment

The shortage of most data sources, and interest in halibut management only during the past several years, has left the Atlantic halibut in a situation similar to that of the Pacific halibut several decades earlier. Management decisions are often a matter of subjective evaluations. Recommendations for Total Allowable Catch (TAC) of Canadian Atlantic halibut have been based largely on recent catch levels. After examining such data in 1987, the Canadian Atlantic Fisheries Scientific Advisory Committee (CAFSAC) Groundfish Subcommittee recommended a precautionary TAC of 3200 t for the Scotian Shelf - southern Newfoundland area, and 300 t for the Gulf of St. Lawrence area. The precautionary TAC's have remained in effect. In reaching the conclusion that these TAC's were required, the Subcommittee stressed that the TAC's reflected neither $F_{0.1}$ nor 2/3 E_{MSY} (effort corresponding to Maximum Sustainable Yield) levels (target levels of fishing mortality employed for those stocks where more extensive data are available), as those values are presently unknown. Further concerns that prompted the Subcommittee to recommend precautionary TAC's were the increased catching power of the recently introduced "circle" hooks compared with J hooks and the increasing value of halibut. The Subcommittee was also concerned about the declining catch rate series (unadjusted for hook type) of halibut longliners (Fig. 27). The relationship of CPUE and exploitable biomass is good, as indicated in the relationship for Pacific halibut (Fig. 28). The concern was accentuated by the fact that the fleet now exclusively uses "circle" hooks, which should cause an increase in CPUE when all other things are

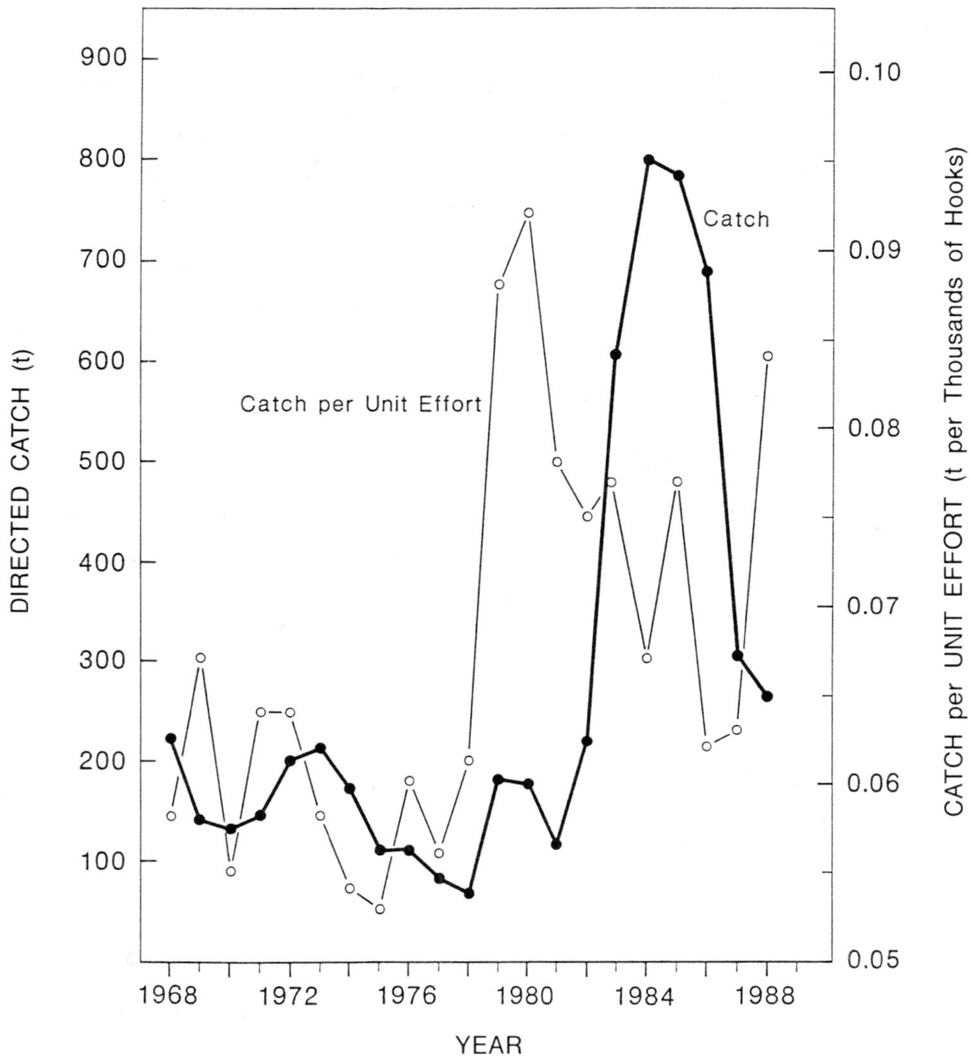

Fig. 27. CPUE series for Scotian Shelf (NAFO Divs. 4VWX) Atlantic halibut taken in "directed" trips by longliners of all tonnage classes.

equal. The Subcommittee chose not to employ a long time series of halibut landings to evaluate the level of the precautionary TAC, as there were indications that landings off Labrador in the 1960's and earlier contained a significant proportion of Greenland halibut.

The IPHC currently estimates CPUE as an index of abundance but uses catch-at-age analysis (Quinn et al. 1985; Deriso et al. 1985) to estimate exploitable biomass of Pacific halibut (Fig. 28). Catch-at-age analysis utilizes the relationship between catch and population parameters by age and year to estimate absolute abundance of year-classes. Weight of catch is directly related to stock biomass after adjusting for changes in fishing mortality, gear selectivity, area by area differences

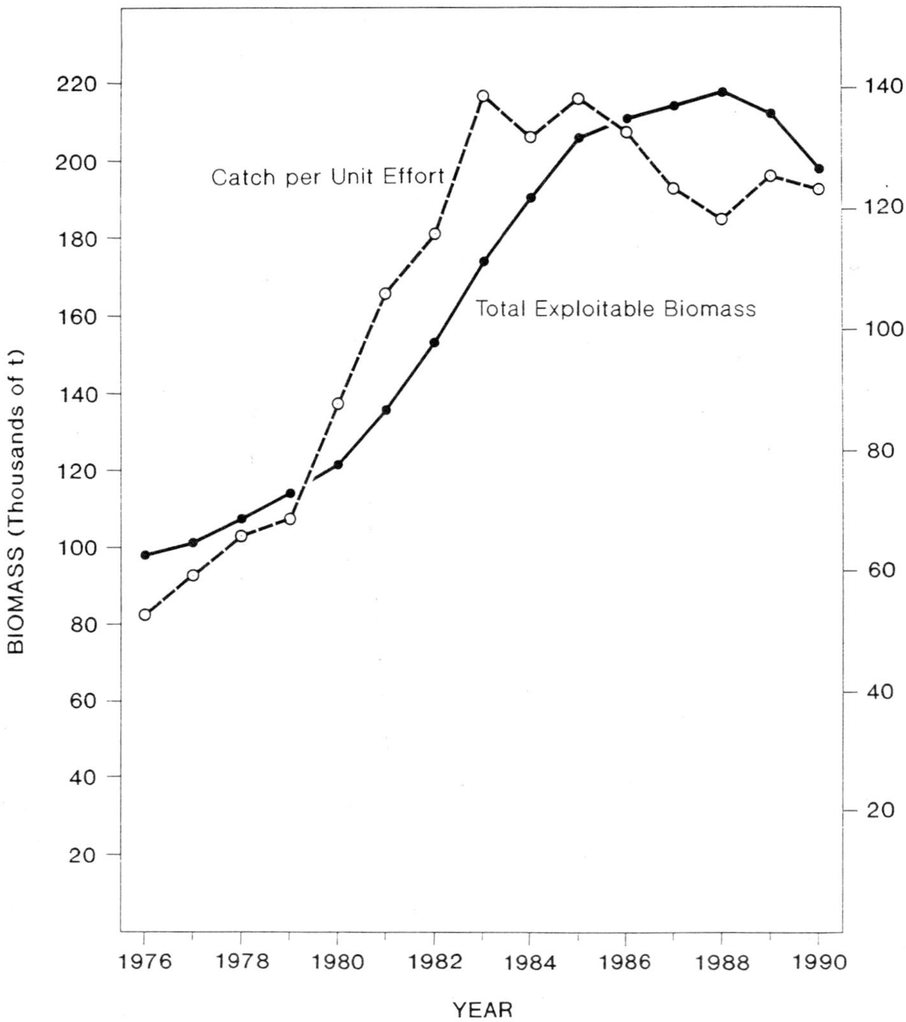

Fig. 28. Exploitable biomass and catch per unit effort for Pacific halibut, 1976–1990 (IPHC, unpublished data).

in catchability, seasonal and regional changes in fishing effort, and fish migration. After adjustments, population abundance at age is related to catch at age over time. Gear selectivity, which accounts for partial recruitment of younger exploitable ages, causes the exploitable biomass to be approximately one-half the total biomass of legal-sized fish. Catch limit recommendations are based on a 35% exploitation rate applied to the exploitable biomass.

Results of Pacific halibut stock assessment through the 1990 fishing year (Patrick Sullivan, IPHC, Seattle, Wash., personal communication) show a continuing coastwide decline in abundance of about 5-6% per year since 1988 to about 142 000 t. The declines are greatest at the extremes of the range, consistent with higher exploitation rates in these areas. Recruitment (abundance of 8-year-old halibut) has declined since a peak in 1985 and is a major factor in the trend of declining abundance.

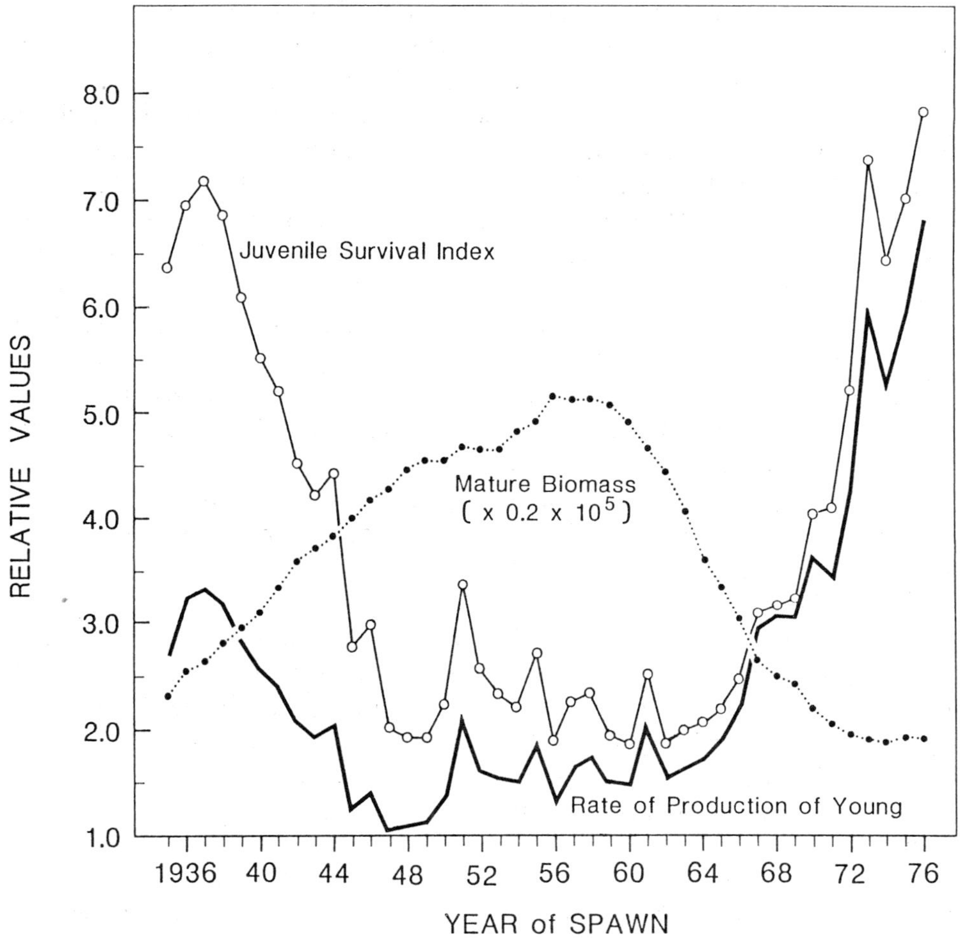

Fig. 29. Juvenile Pacific halibut survival index and rate of production of young are given by year of birth for each year-class. Mature biomass is an index of spawning for each year-class. Units are t for mature biomass, number of 8 year olds per 45.45 kg of mature biomass for juvenile survival index, and 2.2 times biomass of 8 year olds per unit mature biomass for the rate of production of young (from Deriso et al. 1986).

Factors Controlling Production

Debate on the factors controlling production of fish stocks has long raged in the fisheries literature and is a common feature for both Atlantic and Pacific halibut. Most analyses concerned Pacific halibut, dating back to the 1930's, and included density dependence versus independence, and environmental versus fisheries effects. As yet, no definitive conclusions have been reached.

The question of whether fishing or environmental change was responsible for changes in Pacific halibut abundance was a major controversy in the Thompson-Burkenroad debate (reviewed by Skud 1975). Thompson and his supporters claimed that decline in Pacific halibut prior to 1930 and subsequent rebuilding were a result of management restrictions. Burkenroad and his supporters asserted that the environment played the major role. Skud (1975) tended to support Thompson's thesis but pointed out the need for standardization of effort that moderated abundance fluctuations calculated by Thompson.

Schmitt and Skud's (1978) evaluation of spawning biomass and recruitment suggested that recruitment was density dependent. They further suggested that failure of earlier results to show density dependence for Pacific halibut was due to data limitations and to studies that divided the resource into too small components.

More recent analysis of Pacific halibut production focused on the effects of fishing under hypotheses of density dependence or density independence (Deriso et al. 1986). Cyclic high and low values of adult halibut abundance that tend to lag similar cycles of juvenile abundance have been observed for the Pacific halibut. Deriso et al. (1986) proposed two alternative hypotheses to account for the observations: *(i)* juvenile production is density dependent and *(ii)* juvenile production is controlled by unknown exogenous factors.

Under a density dependent hypothesis, Deriso et al. (1986) proposed a stock recruitment relationship through which high adult biomass reduced juvenile production (Fig. 29). In support, they found a statistically significant correlation ($R = -0.87$) of the logarithms of survival against adult biomass while fitting a Ricker spawner-recruit curve. Deriso et al. (1986) indicated that the correlation may be misleading because adult biomass is measured with error and appears in both the dependent and independent variables. They also found a statistically significant negative growth relationship ($R = -0.66$) for adult biomass against average weight of 8 year olds, but noted that 8-year-old weights obtained from commercial catch data may not reflect true sex-specific changes in weight if the sex ratio of the catch changed.

According to the density-independent hypothesis, recruitment fluctuations are induced by intermediate scale environmental cycles. Deriso et al. (1986) observed that juvenile survival, as measured by numbers of 8-year-old Pacific halibut divided by spawning escapement, and juvenile production, measured as biomass of 8 year olds divided by spawning escapement 8 years earlier, exhibited one cycle with peaks in the 1930's and 1970's. Biomass of 8 year olds showed two complete cycles, with peaks in the mid-1940's, mid-1960's, and mid-1980's (IPHC 1988).

The negative effects of increasing fishing mortality (F) on abundance and production were greatest under the density independent hypothesis. A relatively low F within the range analyzed was needed to provide a stable harvest, while higher F drove down abundance during periods of low production. Under density dependence, the population showed less change to increasing F, but yield was higher with a higher F. While highest yield came from high F under the density-dependent hypothesis, the reduction in yield from low F and density dependence was less than the yield lost for high F under the density independent hypothesis.

Parma (1990) demonstrated that under a predictable environment with long term trends, optimal harvest strategy depends on management objectives. An objective to maximize yield (a risk-neutral strategy) would result in increased spawning escapement during good conditions, but decreased escapement during poor conditions. This strategy provides a feedback leading to a boom-bust cycle. An objective for stable yield (a risk-averse strategy) tends to less variable escapement. Parma and Deriso (1990) analyzed the problem of managing a fish population under competing hypotheses of density dependence and density independence and compared optimal risk-neutral, risk-averse, and constant harvest rate strategies using simple representations of the problem. The optimal strategy changes depending on which hypothesis is deemed correct. Actively adaptive policies call for deliberate experimentation with spawning quotas to reduce uncertainty, while passively adaptive policies provide only passive learning, because the effect of learning from alternative management actions is ignored. Some combinations of spawning escapement and harvest result in similar stock response for both hypotheses and provide no useful information for evaluating the alternative hypotheses. Optimal risk-neutral policies, which provide the greatest long term yield, resulted in high variability in annual harvest; yields were only slightly higher than those obtained from the less variable risk-averse policies. Constant harvest rates could potentially result in substantial loss of yield unless the harvest rates were updated periodically using information gained actively or passively to evaluate the probability of the alternative hypotheses. Such updating is currently underway for Pacific halibut.

Sutcliffe (1973) correlated landings of Atlantic halibut in the Gulf of St. Lawrence with freshwater discharge from the St. Lawrence River. He found a significant relationship (R = 0.855) when the Quebec catch was compared with discharge during the month of March (moving averages of 3 years were employed, with discharge lagged 10 years). However, the analysis made no attempt to adjust for effort nor was any mechanism proposed for the linkage. Moreover, a recent reexamination of the hypotheses (Drinkwater 1987) indicated that the correlation has not held up well during more recent years.

The latitudinal ranges of North American Atlantic and Pacific halibut (40-60°N) and their centers of distribution on the continental shelf are similar on either coast, yet the production of the two species is substantially different, with Pacific halibut about an order of magnitude higher. The current low levels of Atlantic halibut production are probably the result of past overexploitation. The similar life histories, especially age, growth, fecundity, and maturity, suggest no production impediments for Atlantic halibut.

Fisheries Management

History of Regulation

Management practices for Atlantic and Pacific halibut diverged early in the history of the North American fisheries. Pacific halibut fishing has operated under a single international management body since 1923, with increased management complexity and sophistication since then. Regulations for areas, catch limits, gear restrictions, and size limits have improved based on one of the longest running data sets in fisheries. Halibut was treated as part of the general groundfish resource in the Atlantic, so did not experience single species management until 1988, when requested by the Canadian industry. Provisional regulations were implemented based on best available, but very limited, data. No specific regulations for Atlantic halibut are in force by the U.S.

Pacific Halibut

The International Pacific Halibut Commission was formed in 1923, largely resulting from industry efforts to combat declining resource abundance. Recognizing that they could not regulate themselves, fishermen and processors petitioned the governments of the U.S. and Canada to provide for joint management. The original treaty that formed the Commission, the Halibut Convention of 1923, was renegotiated in 1930, 1937, and 1953, and revised by protocol in 1979. During these years, the authority and responsibility of the IPHC has increased; the Commission is currently directed to maintain Pacific halibut stocks at levels that permit the optimum yield from the fishery and may do the following.

1. Divide the Convention waters into regulatory areas.

2. Establish one or more open or closed seasons for each area.

3. Limit the size of the fish and the quantity of the catch to be taken from each area within any season during which fishing is allowed.

4. During both open and closed seasons, permit, limit, regulate, or prohibit the incidental catch of halibut that may be taken, retained, possessed, or landed from each area or portion of an area, by vessels fishing for other species of fish.

5. Fix the size and character of halibut fishing appliances to be used in any area.

6. Make such regulations for the licensing of vessels and for the collection of statistics on the catch of halibut as it shall find necessary to determine the condition and trend of the halibut fishery and to carry out the other provisions of this Convention.

7. Prohibit halibut fishing in any area or portion of an area that the Commission finds to be populated by small, immature halibut and designates as nursery grounds.

The Protocol of 1979 also provided for regulations of recreational fishing for Pacific halibut and for a prohibition of nationals from either country fishing for Pacific halibut in waters of the other country.

While Pacific halibut are considered a single stock because of intermingling during egg and larval drift, the Pacific region under IPHC authority is divided into management regions to distribute the catch. Adult halibut appear to migrate only small distances on the fishing grounds from summer to summer, so could experience local depletion in specific areas if a single catch limit were applied to the entire area. Counter-migration by juveniles could eventually replenish over-fished areas but only after a lag of several to many years if fishing pressure were reduced.

While halibut are caught by a variety of fishing gear, Pacific halibut may be retained only by hook and line gear and only during open seasons. The fishery developed in Pacific waters with longline and handline gear; troll gear entered the fishery as salmon trollers caught halibut incidentally. Trawl gear was first attempted, unsuccessfully, for Pacific halibut fishing in 1906 by two vessels that subsequently converted to dory fishing (Bell 1981). Low catches and small halibut made the ventures unprofitable. Several other attempts over the next 15 years also failed. Trawl retention of Pacific halibut was prohibited in 1944 (Bell 1981) after a trawler successfully targeted halibut in 1943, and other trawlers had been retaining halibut caught incidentally to groundfish fishing. Justification involved the tendency of trawls to catch small halibut below the legal size and the calculation that yield of halibut would be greater by allowing small halibut to grow to larger size. Setnets were prohibited in the Pacific halibut fishery in 1938 to prevent their imminent use (Bell 1981), after experience in Norway (Devold 1938) convinced IPHC that setnets were undesirably selective for large spawners.

During the biomass increase since 1974, the philosophy and techniques for management and stock assessment for Pacific halibut changed. In the early 1980's, IPHC adopted a management philosophy of reserving 25% of the ASP, the biomass produced in a year above that necessary to keep the abundance constant, to help stock rebuilding. The remaining 75% of the ASP was utilized for the directed fishery.

In 1985, when the halibut resource seemed to be rebuilt, IPHC reduced emphasis on ASP management and shifted to a Constant Exploitation Yield (CEY) management. An optimal exploitation rate of 35% was chosen in 1986 to calculate yield from exploitable biomass. The IPHC adopted a policy that harvest in regulatory areas should occur in proportion to exploitable biomass in the regulatory areas.

Atlantic Halibut

Prior to 1988, the Canadian Atlantic halibut fishery went unregulated. As a result of increased interest in the fishery and concerns expressed by components of the fishing industry, the Assistant Deputy Minister, Atlantic Fisheries, formed a special working group on Atlantic halibut under the auspices of the Atlantic Groundfish Advisory Committee. The mandate of the group was to advise on appropriate management measures to regulate fishing on Atlantic halibut. As a result of its initial deliberations, the Working Group referred the following three questions for consideration to the Canadian Atlantic Fisheries Scientific Advisory Committee (CAFSAC), the body responsible for the provision of scientific advice for management of Atlantic fisheries.

1. What should, on a biological basis, be the management unit(s) for Atlantic halibut on the Atlantic coast?

2. Can CAFSAC advise on the total allowable catch for the management units for the 1988 season?

3. What is the biological impact of the different gear used in the halibut fishery ("circle" hooks, regular hooks, and otter trawls)?

In response to a request from CAFSAC, Neilson et al. (1987), using tagging results (Bowering 1986 and Stobo et al. 1987) and age and growth comparisons (Bowering 1986), concluded that the Atlantic halibut of the Scotian Shelf and southern Newfoundland banks should be treated as a single unit. The data also indicated that a separate stock occurs in the Gulf of St. Lawrence.

These two management units were put into place as recommended. The current interpretation of stock structure has been supported by a parasitological study of Atlantic halibut by Scott and Bray (1989) who noted geographic changes in parasite prevalence and intensity in the form of clines. Such differences may be explained by gradual changes in environment and gave no indication of stock separation throughout the Scotian Shelf. No specimens from the Grand Banks area were included in the study.

Insufficient data were available to perform any type of analytical assessment of TAC. Therefore, based largely on recent catch levels precautionary TAC's of 3200 t for the Scotian Shelf - southern Newfoundland area and 300 t for the Gulf of St. Lawrence were introduced for 1988. It was stressed by CAFSAC that these levels did not reflect either $F_{0.1}$ or 2/3 E_{msy} normally employed for exploitation of Canadian groundfish stocks but were simply preemptive ceilings on catch until sufficient scientific information becomes available to assess these resources adequately.

Sufficient data were also unavailable to fully evaluate the biological impact of different gears used in the fishery. It was noted, however, that the circle hook was at least twice as efficient at catching Atlantic halibut compared with the J hook (Peeling and Rogers 1984; Cliche 1986).

As a result of further pressures from some components of the fishing industry, a minimum size limit of 32 in. (81.3 cm) was also introduced for 1988. This measure will be discussed in detail in the following section.

Management Issues

Several significant management issues have evolved that have counterparts in both the Atlantic and Pacific fisheries. Some of these have been resolved, whereas others remain. Three of the most significant are minimum size, bycatch, and new entrants into the fisheries.

Minimum Size Limits

Both Atlantic and Pacific halibut are managed with a 32 in. (81.3 cm) commercial size limit, but the timing and rational for the limits took different paths for the two fish. Size limits have been a feature of the Pacific fishery since 1940. One change to the limit, from 24 to 32 in. (65 to 81.3 cm), occurred in 1973 to maximize yield per recruit (Myhre 1974). The Atlantic size limit implemented in 1988 maximized value, rather than yield, from the resource (Neilson and Bowering 1989).

The management measure of a minimum fish size was first introduced on the Pacific Coast in 1940. The regulation then stated that, "The catch of halibut to be taken from each area during the halibut fishing season of 1940 shall also be limited to halibut weighing 5 pounds or over as computed with heads off, entrails removed or to halibut weight 5 pounds 13 ounces or over as computed with heads on, entrails removed and the possession of any halibut of less than the above weights by any vessel or by any master or operator of any vessel or by any person, firm or corporation, is prohibited" (Pacific Halibut Fishery Regulations 1940). Interestingly, the IPHC did not have the mandate to regulate minimum landed size under the 1930 Convention. That authority was forthcoming in the revision to the Convention completed in 1953.

In 1944, the minimum size regulation was rewritten to include a measure of length as well as weight. The regulation then stated, "The catch of halibut to be taken from each area during the halibut fishing season of the year 1944 shall also be limited to halibut which with the head on are 26 inches or more in length as measured from the tip of the lower jaw to the extreme end of the middle of the tail or to halibut which with the head off and entrails removed are 5 pounds or more in weight, and the possession of any halibut of less than the above length or the above weight, according to whether the head is on or off, by any vessel or by any master or operator of any vessel or by any person, firm or corporation, is prohibited."

The biological basis for the initial regulations was never documented. The choice of the minimum size of 5 lbs. likely reflected the division between the "baby" and "chicken" market categories. Because few vessels landed "baby" halibut, the regulation probably had little effect on the fishery (Myhre 1974). It is also worthwhile noting that the regulations were implemented during World War II, when the activities of the Commission and those of the fishery in general were in a hiatus. However, the implementation of a 32 in. (81.3 cm) size limit was given careful study (Myhre 1974). Myhre used a modified Ricker yield model to show that the sustainable yield could be increased by raising the size limit, provided that the mortality of undersized fish caught and released was not too high. Hoag (1975) subsequently addressed that problem and concluded from tag returns that about 50% of trawl-caught Pacific halibut less than 80 cm survived if release occurred within 15-20 min.

The Atlantic coast groundfish fishery is dominated by otter trawl gear, and in contrast with the Pacific situation, retention of trawl-caught halibut has been permitted, though since 1988 only fish larger than 32 in. (81.3 cm) are legal. Fish caught by bottom trawlers tend to be smaller, on average, than those caught by longline gear (Fig. 18). Until recently, there has been little appreciation of the extent of mortality of such incidentally caught fish, which may be on the order of 65% (Neilson et al. 1989).

The effect of the minimum size regulation in the Canadian Atlantic halibut fishery was assessed from the point of view of yield and value per recruit (Neilson and Bowering 1989). The model indicated that under most scenarios, the imposition of the size limit would not result in increased yield per recruit (Fig. 30). In general, yield per recruit was more sensitive to fishing mortality than age of first entry to the fishery. While reduced yields were usually associated with the minimum size limit, the minimum size regulation appeared to be an attractive option when value was considered, reflecting the higher value of the larger market categories (Fig. 30). Both yield and value per recruit were found to be sensitive to the choice of the natural mortality rate.

Enforcement of minimum size regulations is always a problem. Incidentally caught halibut are usually not very numerous, and the occasional halibut is traditionally viewed as a "prize," which is often retained for home consumption. Another significant problem related to bycatch is the wasteful practice of landing immature halibut (<60 cm) mixed with the other species of flatfish common in the Atlantic area (including winter flounder (*Pseudopleuronectes americanus*), yellowtail flounder (*Limanda ferruginea*), American plaice (*Hippoglossoides platessoides*), and witch flounder (*Glyptocephalus cynoglossus*)). The mixed landings are marketed generically as "flounder" and typically fetch a lower unit price than the smallest legal market categories of halibut. Such problems can be overcome by clearly demonstrating that fishermen can expect increased value for their landings if undersized fish are returned to the water. Alternatively, discard of undersized halibut could be made mandatory, as is the case for the Pacific halibut, and enforced by inspection of catch at sea or in the fish plants.

Bycatch

Atlantic and Pacific halibut are caught inadvertently in fisheries targeting on various groundfish and shellfish species, and indications of bycatch mortality are substantial. No bycatch of Pacific halibut may be retained, except by hook and line during open fishing periods. Atlantic halibut of all sizes had traditionally been landed with groundfish catch, until the minimum size limit restricted landings to legal-sized bycatch. Management problems are different, as groundfish fisheries off Alaska are restrained by Pacific halibut bycatch, yet the Atlantic halibut longline fishery in some

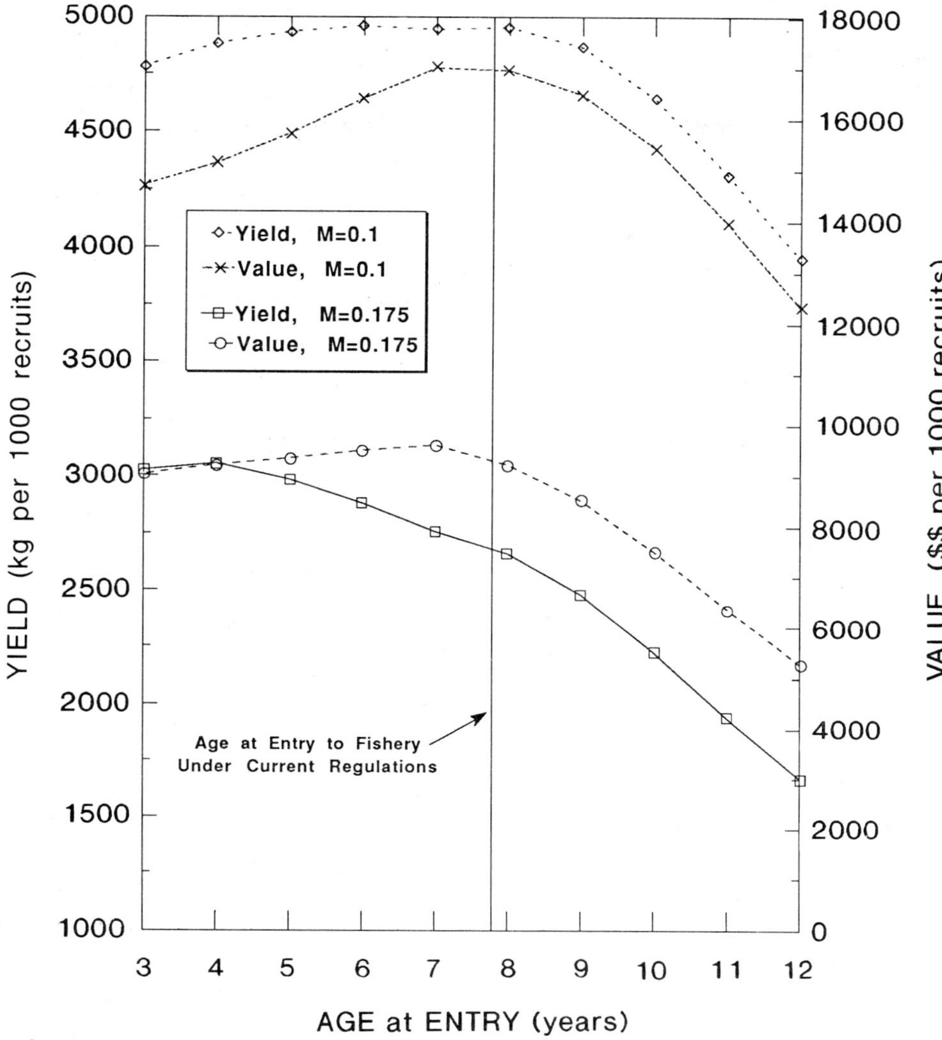

Fig. 30. Yield and value per recruit versus age of entry to the Atlantic halibut fishery. Natural mortality (*M*) was either 0.1 or 0.175, fishing mortality (*F*) was assumed to be 0.3, and the selection curve flat topped (from Neilson and Bowering 1989).

areas is constrained by bycatch of Atlantic cod. The Pacific halibut fishery catch limits are reduced to compensate for lost yield caused by bycatch mortality, but Atlantic halibut TAC's are not adjusted for bycatch impacts, as estimates of bycatch mortality are not made.

The U.S. National Marine Fisheries Service (NMFS) supplied the IPHC with estimates of the incidental catch in foreign and joint venture groundfish fisheries through an observer program. The NMFS has also supplied bycatch data for the fully domestic groundfish fishery since a mandatory observer program began in 1990. Estimates of bycatch in other fisheries or in the fully domestic groundfish fishery before 1990 were generated by IPHC and NMFS staffs from information collected on research surveys or through predictive models (Williams et al. 1989). Landed bycatch of Atlantic halibut is accounted for by fish tickets, but no estimates are made of discarded bycatch. The IPHC has the authority to prohibit retention of Pacific halibut in groundfish fisheries but cannot regulate the groundfish fisheries to moderate halibut bycatch. Retention of Atlantic halibut in Canada is managed as part of the groundfish fishery, and no Atlantic halibut bycatch management occurs in U.S. waters.

Most halibut that are incidentally caught are injured to some degree during capture. However, not all fish that are returned to the sea die, so the incidental mortality is less than the actual catch. The likelihood of a halibut being killed during incidental capture depends on the fishing operation (Hoag 1975; Neilson et al. 1989). Both Hoag (1975) and Neilson et al. (1989) present data indicating increased mortality of trawl-caught halibut as time on deck increased, as catch increased, and for smaller fish.

Hoag (1975) used relative tag return rates for halibut classified by condition factor (physical status of released fish) to estimate mortality of halibut discarded as bicatch from trawlers. Williams et al. (1989) summarized the assumed discard mortality rates for gears used in fisheries that cause Pacific halibut bycatch, based on fishery observers' conditioned factor data and Hoag's (1975) bycatch survival estimates for conditioned factors. Mortality in trawls with long tows, large catches, and slow sorting is usually very high, approaching 100%. Trawling operations that transfer trawl codends to a mothership for processing also exhibit discard mortality rates close to 100%, as the sorting process is very slow and the catches are usually large. Discard mortality in short trawl tows with small catches and quick sorting has been estimated at 50%. Discard mortality in crab pots is believed to be 10%. Lack of observers on fully domestic vessels prior to 1990 required application of foreign and joint venture data to the domestic fleet. Gregg H. Williams (IPHC, Seattle, Wash.) and Thomas K. Wilderbuer (NMFS, Seattle, Wash.) reestimated discard mortality rates, using observer data from 1990, as 65% for trawlers in the Gulf of Alaska,75% for trawlers in the Bering Sea - Aleutian Islands, 16% for longline in both areas, and 10% for groundfish pots in both areas. Use by the longline fisheries for sablefish (*Anoplopoma fimbria*) and Pacific cod (*Gadus macrocephalus*) of devices that automatically strip fish from the hooks increases the discard mortality rate of Pacific halibut for longline fisheries by a factor of about 3.5 (Stephen M. Kaimmer, IPHC, Seattle, personal communication).

Neilson et al. (1989) examined the survival of sublegal sized Atlantic halibut caught in longline and bottom trawl gear using live holding facilities on board a research vessel and subsequently, in a land-based laboratory. Commercial practices were simulated during fishing operations. Of halibut less than the proposed size limit, 35% of the otter trawl catch and 77% of the longline catch survived more than 48 h. Such rates of survival are consistent with, but somewhat lower than, those observed in the Pacific fishery. Factors potentially influencing halibut survival (handling time, total catch, fish length, maximum depth fished, and trawl duration) were examined using proportional hazard models. On the basis of those analyses, it was concluded that in bottom trawl sets of similar duration used in the commercial fishery (2 h), higher survival times were associated with shorter handling time, larger fish size, and comparatively small total catch weight. Supplemental information on the condition of trawl-caught halibut was also obtained from observers stationed on board commercial trawlers. Interestingly, using criteria similar to those employed by Hoag (1975), observers indicated that about 75% of trawl-caught halibut were in excellent condition. Yet during the course of the experimental fishing, halibut that appeared to be in excellent condition often succumbed later while in the holding tanks. Neilson et al. concluded that using Hoag's condition factor overestimated survival. The IPHC staff (Seattle, Wash., unpublished) has observed variable mortality of tank-held fish and concluded that holding halibut in tanks may cause mortality not experienced by halibut released to the sea. However, tank-holding mortalities demonstrate that employing visual assessment of condition as an indicator of future survival should be employed cautiously.

Pacific halibut bycatch management has historically emphasized control of foreign groundfish fishing fleets, which accounted for nearly all groundfish harvest and halibut bycatch mortality from the 1960's through the 1980's. Restrictions evolved over time, usually involving time and area closures, through bilateral negotiations with foreign countries and through the North Pacific Fishery Management Council (NPFMC) after passage of extended jurisdiction (Fredin 1987; Freeberg, in press). While no analysis has shown time and area restrictions of the foreign fleets to be effective, no consistent decrease in bycatch rates occurred during this period (Trumble 1992). The NPFMC did achieve substantial reduction in halibut bycatch rates from 1985 when it mandated a 50% reduction in bycatch rates of foreign fleets over a 5-year period. The Japanese, which made up most of the foreign fishing at the time, allocated bycatch to fishing companies and then to individual vessels. The 50% rate reduction was achieved in 3 years. Figure 26 shows the reduction in estimated halibut bycatch mortality through 1985.

Since 1986, increase in Pacific halibut incidental mortality has occurred despite a reduction in foreign fishing off Alaska and is attributed to increased mortality by joint venture and fully domestic groundfish fisheries in the U.S and Canada. Groundfish fishing in Alaska accounts for 85 to 90% of the Pacific halibut bycatch mortality, with Alaskan crab and shrimp fishing and Canadian groundfish fishing contributing the remainder. The objectives of U.S. extended jurisdiction legislation in 1976 included fully "Americanizing" the groundfish fisheries in U.S. waters. The NPFMC adopted policies and regulations that encouraged joint ventures between domestic catcher vessels and foreign processors when U.S. processing capacity was insufficient. As domestic processing increased, both through at-sea processing vessels and shore-side plants, harvest priorities shifted to fully domestic operations. In 1990, the full harvest of groundfish was allocated to the fully domestic fleet. As the mix of foreign, joint venture and fully domestic fishing changed over the past several years, so did the source of bycatch mortality (Fig. 31). The NMFS provided estimates of foreign and joint venture bycatch from a comprehensive observer program required by the Magnuson Fishery Conservation and Management Act. No such requirements applied to the fully domestic U.S. groundfish fishery until 1990. Bycatch estimates were made indirectly from estimated rates based on joint

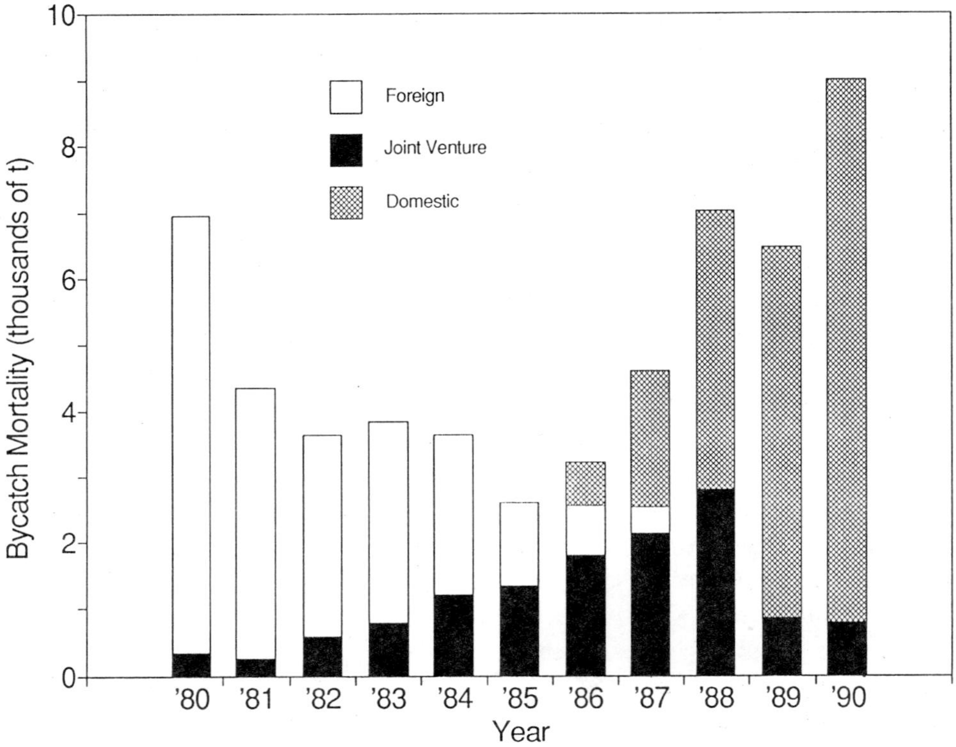

Fig. 31. Estimated mortality of Pacific halibut taken as bycatch in the groundfish fisheries of Alaska, 1980–1990.

venture or research bycatch rates. Current mandatory observer coverage provides direct estimates. Observer coverage is not required in the Canadian Pacific groundfish fishery, but planning is underway for an observer program.

It is very difficult to make precise estimates of the effects of bycatch on the commercial-sized component of halibut stocks because bycatch is largely made up of younger migrating halibut. Growth, mortality, and migration greatly complicate the estimation procedures. If the same age composition occurred in both fisheries one could consider the bycatch removals as merely increasing the directed removals. Migration rates of juvenile halibut are not well known, so the impact of bycatch of juvenile halibut from specific areas on adult populations in those or other areas must be made estimated indirectly.

Bycaught halibut are generally smaller than those harvested by the directed fishery. Consequently, factors such as maturity, reproductive capacity, survivorship, and growth substantially affect stock productivity. By allowing small halibut to remain at large for a longer period of time, a net gain in stock biomass occurs owing to the greater cumulative gain in individual weight relative to losses incurred owing to mortality. Smaller fish are less likely to be mature and have less reproductive capacity. Those harvested earlier in their life history not only contribute less in terms of short-term yield, but they also contribute less to the maintenance of future stock biomass or to future yields. Bycatch losses affect recruitment, future catch, and future reproductive potential of the stock.

The IPHC's approach for compensating the stock for bycatch losses is to reduce harvest in the directed fishery such that the reproductive potential of the exploitable component of the stock would be the same after bycatch as it would have been if bycatch had not occurred (Sullivan et al. 1993). The compensation factor was determined to be 1 t of catch limit reduction for each metric ton of bycatch mortality. If the reproductive compensation was done correctly and if the bycatch is estimated correctly, the halibut spawning stock will remain in the same condition whether bycatch occurs or not. The directed halibut fishery pays for maintenance of the resource through lower catches. Catch limit reduction occurs the year after the bycatch, while spawning from the bycaught fish would have occurred over several years in the future. While timing of egg replacement is not exactly correct, improvements would require accurate forecasts of future recruitment, which we cannot yet do. The IPHC approach, to our knowledge, is unique in its treatment of bycatch compensation.

Impact on the directed halibut fishery consists of two parts: the catch limit reduction to maintain reproduction and reduced recruitment to the directed halibut fishery from bycatch of pre-recruits.

Reproductive compensation for bycatch immediately deprives the directed fishery of 1 t of yield for each metric ton of bycatch the previous year. However, this amounts to leaving fish in the stock rather than catching them right away, and some are caught later. On average, about 0.6 of the 1-t bycatch compensation is eventually caught, so the net impact of reproductive compensation is 0.4 t / t bycatch.

Bycatch eventually reduces recruitment to the directed fishery and amounts to 1.2 t lost yield / t bycatch.

The combined effects of reproductive compensation and lost recruitment shows a net loss to the directed fishery of 1.6 t / t bycatch: 0.4 t from reproduction compensation and 1.2 t from reduced recruitment.

Although the otter trawl component of the fishing industry is responsible for a substantial proportion of the Atlantic halibut landings, it is recognized that the catches are generally undirected and are bycatches of other major groundfish fisheries such as cod. As a result, mobile gear vessels are restricted to a 5% bycatch limitation in its directed groundfish fisheries, and excess bycatch must be discarded. That catch of Atlantic halibut is relatively low compared with the catch of the directed species, but the proportion of the Atlantic halibut population taken as bycatch is unclear. For the directed longline fishery for Atlantic halibut, on the other hand, there can be significant bycatches of other groundfish, particularly cod. For most areas now under regulation this is not a problem, with the exception of the southern Newfoundland Grand Bank where the bycatch of cod often exceeds the directed catch of halibut. To alleviate this problem, this directed Atlantic halibut fishery has been allocated a portion of the TAC for cod in the same area. The fishery for Atlantic halibut is then closed when either the cod or Atlantic halibut allocation is taken, whichever comes first, and the Atlantic halibut TAC in the southern Grand Bank area may not be taken.

Entrants into the Fishery

The U.S. and Canada have taken very different approaches to management of new entrants to the halibut fisheries. Canada implemented license limitations for its Pacific halibut fishermen following extended jurisdiction in 1977; fishermen not meeting qualifying standards of minimum landings were excluded from the fishery. In 1989, Canadian fishermen requested a share quota system, which was implemented in 1991. Open access remains in effect for Atlantic halibut, although Canada allocates to gear and vessel size categories. Current lack of Atlantic halibut in U.S. waters precludes need for regulation. Open access continues in the U.S. Pacific halibut fishery, although the NPFMC approved share quotas for individual fishermen that are scheduled for 1995.

Once the upper limits for harvesting Atlantic halibut were set, the TAC's among various fleet sectors were allocated on the basis of the previous 5 years' catch history. The fleet components to which allocations were given are as follows.

1. Fixed gear vessels less than 19.8 m.
2. Mobile gear vessels less than 19.8 m.
3. Fixed gear vessels 19.8-30.5 m.
4. Mobile gear vessels 19.8-30.5 m.
5. Fixed gear vessels greater than 30.5 m.
6. Mobile gear vessels greater than 30.5 m.

For these fleet sectors, only fixed gear components were permitted to have a directed fishery for Atlantic halibut, whereas the mobile fleet components were permitted allocations strictly as a bycatch of other groundfish fisheries. Although an overall TAC was set for the Scotian Shelf - southern Newfoundland Banks area, the allocations by fleet sectors were further subdivided three ways among the Scotian Shelf, St. Pierre Bank, and the southern Newfoundland Grand Bank.

Success of rebuilding the Pacific halibut resource during the 1980's, decline of the red king crab (*Paralithodes camtschatica*) in 1982, and development of limited entry in the salmon fisheries led to increased effort and increased catching capacity in the Pacific halibut fishery during the past decade. As a comparison, the 1987 catch limit of about 30 000 t in the Gulf of Alaska was taken in three fishing periods of 24 h each, whereas over 40 days were required to take annual catch limits of about 12 000 t during the late 1970's. Gulf of Alaska harvest currently represents about 70% of the total coastwide halibut harvest and the volume of product landed in very short periods overloads processing capacity.

The short seasons are a result of increased fishing effort and higher catch rates in spite of increased catch limits (Fig. 32). Catch rates have increased owing to higher abundance and improved fishing techniques such as circle hooks and new navigation equipment. New vessels have entered the halibut fishery because of the high catch rates and relatively high halibut prices in recent years, and because halibut

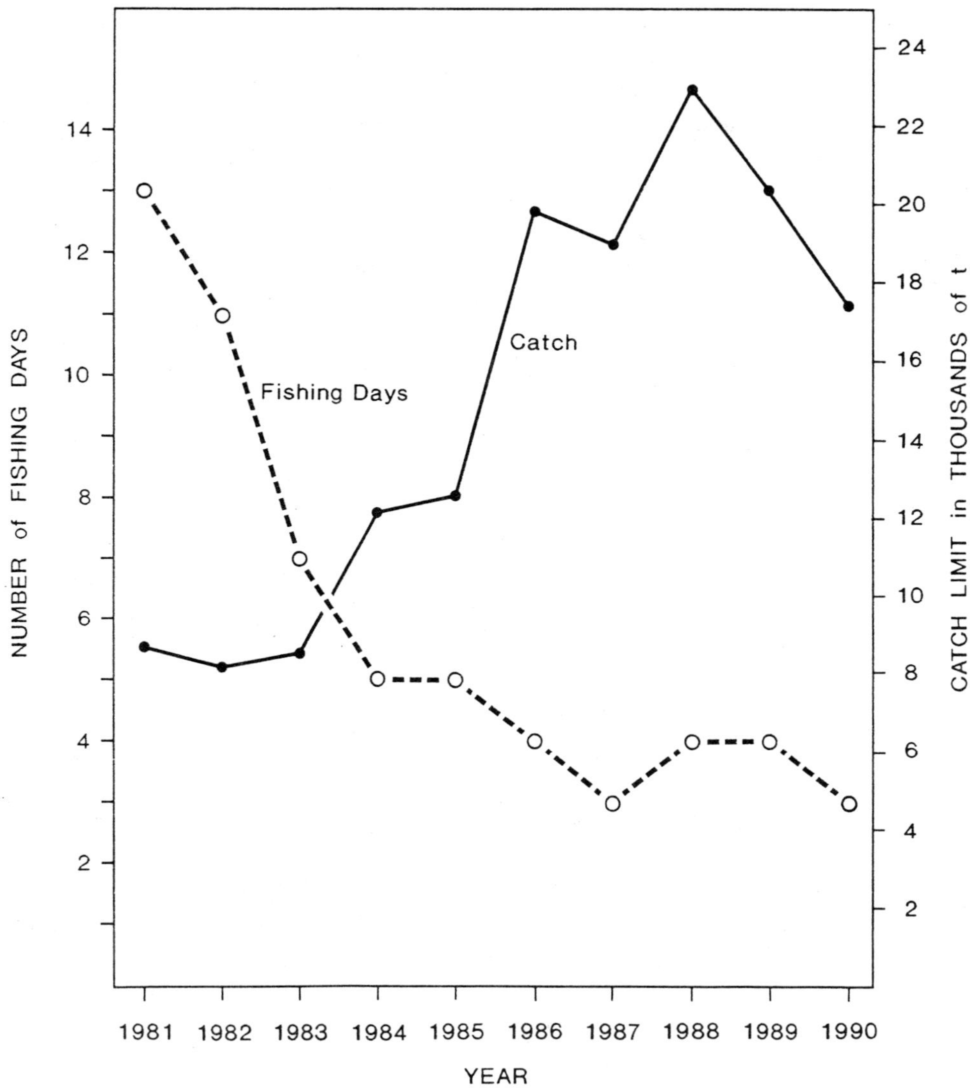

Fig. 32. Changes in catch limits and fishing days for Pacific halibut in IPHC Area 3A (central Gulf of Alaska), 1981–1990.

longline fishing gear is relatively inexpensive. The short intense fishery has created the following management problems.

1. It is no longer possible to closely attain the catch limit with unrestricted 1-day openings because of the high daily catch rates. For example, 16 300 t were landed in the central Gulf of Alaska in two 24-h fishing periods during 1987, 2400 t below the 18 700 t catch limit. The high risk of greatly over or under fishing requires that restrictions be implemented, at least for the last fishing period.

2. Very intense fishing during short fishing periods causes much of the halibut to be frozen rather than reaching the fresh market and reduces the quality of halibut landed.

3. Some 1000 to 2000 t of halibut are wasted annually as a result of lost or abandoned gear, poor handling of halibut below the minimum size limit, and exploration by fishermen for halibut under the guise of fishing for other species prior to each fishing period. Although some wastage is inevitable in any fishery, the short intense fishing periods have increased the problem. Multiple short openings present more opportunity for wastage than do fewer openings conducted at a slower pace.

4. Increasing reports of fishing before and after an opening suggest that the 1- or 2-day seasons create greater enforcement problems. The IPHC staff is concerned that violations will become increasingly widespread unless enforcement effort is increased or alternative restrictions are implemented.

5. The short fishing periods create potential safety problems because of economic pressure to fish during poor weather conditions and to haul as much gear as possible. For example, at least two vessels and five lives were lost during the 1988 fishery off Alaska.

Two general methods can be used to address these problems. The first is an access limitation program and the second is efficiency reduction. The IPHC does not have authority to implement limited access and has chosen not to develop efficiency reductions. Authority for limited access decisions is given to the NPFMC for waters off Alaska. The NPFMC developed a moratorium on new licenses for Pacific halibut in 1983 while a long term limited access program was being studied, but opposition and political pressure from fishermen caused the U.S. government to reject the program. The NPFMC started a new study of limited access for Pacific halibut in 1989 after several years of discussing various options for access limitation for fisheries in general. An individual quota system is scheduled to be in place for 1995.

Future of the Resource and Fisheries

Atlantic and Pacific halibut face the future from substantially different starting points. Pacific halibut have benefitted from fishery management that allowed the resource to take advantage of an opportunity to expand to near record abundance during the mid-1980's, although natural declines have occurred since. The IPHC estimates that a sustainable harvest of 25 000 to 40 000 t is possible into the future. Atlantic halibut are in an uncertain condition but are probably at a diminished level following a history without fishery management. Data are insufficient to predict

Atlantic halibut harvest other than within the historical range of 2000-5000 t, and stock assessment analysis cannot yet predict potential productivity of the resource.

Atlantic

In recent years many Atlantic groundfish stocks have experienced significant declines in TAC's as a result of declining stock sizes. As one might expect, these reductions seriously impact many components of the fishing industry, particularly those that are already overcapitalized. In an attempt to alleviate this problem, the industry is exploring the possibility of diverting effort to underutilized species such as silver hake, Greenland halibut (turbot), and grenadiers. Some of these affected parties consider that Atlantic halibut TAC's are unnecessarily restrictive. Indeed, there was considerable opposition to introducing any TAC's on Atlantic halibut until justified by a more quantitative scientific assessment of the resource. The effect of reduced TAC's on other groundfish can also be felt by certain components of the existing Atlantic halibut fishery such as in the southern Newfoundland Grand Bank area. Significant allocation of cod is given to the Atlantic halibut fleet because of its large bycatch of cod. With a reduction in the TAC for cod in this area from 40 000 t in 1988 to 25 000 t in 1989 so came a proportional reduction in the allocation to the Atlantic halibut fleet. It is conceivable, therefore, that the cod bycatch allocation could be reached before the Atlantic halibut TAC is taken. Northern cod stocks, which were severely depressed and closed to fishing in the early 1990's, have little interaction with Atlantic halibut through bycatch.

Pacific

The future of the Pacific halibut fishery depends on two factors: condition of the resource and the management system put into place. Sufficient knowledge appears to be available to maintain the resource at highly productive levels well into the future. The abundance fluctuates over time, possibly in a cyclic pattern, through changes in recruitment, productivity, and harvest. Although management cannot control natural fluctuations, it can control the total fishery-induced mortality. As abundance changes, the management structure can respond with appropriate catch limits.

The ability to respond, however, does not guarantee that response. The future of the fishery depends on the specifics of whatever management system is applied to Pacific halibut, and the importance that Pacific halibut may play in the overall management of fishery resources. Since the peak exploitable abundance in 1986, the Pacific halibut resource has declined about 5-10% per year. Projections for continued declining recruitment lead to a forecast of annual reductions in exploitable biomass perhaps to the mid 1990's.

Future harvest from the directed fishery could easily fall within 20 000 to 35 000 t if commitment to the fishery continues. Conversely, the directed harvests could fall to low levels if the halibut resource is diverted to such uses as bycatch in groundfish fisheries. Commercial fisheries for the large groundfish resource (2.3 million t TAC in 1990) in Alaska waters has been a major source of halibut bycatch. Continuing demand for halibut as bycatch has increased estimated coastwide bycatch mortality from a recent low of 4400 t (7000 t of yield loss) in 1985 to 10 500 t (16 800 t of yield loss) in 1990. The bottom trawl fisheries have accounted for most of this bycatch. A longline fleet targeting on Pacific cod is rapidly growing, with a concomitant increase in halibut bycatch, and longline fisheries for other groundfish species, especially sablefish, add to bycatch losses.

As increased bycatch puts increased pressure on the directed Pacific halibut fishery, so increased catching capacity in the Pacific halibut fishing fleet puts pressure on the commercial fishing seasons. Under open access, short, 1-day fishing seasons will continue in the high production areas off Alaska. One of the biggest controversies in management of Alaska fisheries is the future of open access for halibut and groundfish fisheries and whether license limitation or individual quotas will be adopted. Without a limited access program that effectively reduces the Pacific halibut fleet, little change in the harvesting regime will occur, and the management problems identified earlier will continue. Action by the NPFMC to begin allocating halibut quota to individual fishermen is intended to help solve them.

Quality of Scientific Information

As expected from the long history of Pacific halibut management relative to that in the western Atlantic, scientific data are far superior in the Pacific. However, lack of scientific information upon which to base the development of the fisheries has hampered the development of the halibut fisheries to some extent on both sides of North America. As can be seen from Table 5, the shortage of scientific data for adequate management of the Atlantic fishery is more acute than the Pacific fishery. Comparatively few baseline biological studies have been completed on the Atlantic coast, with particular shortcomings in early life history and detailed stock-related investigations. Of even greater concern from a management perspective is the paucity of population parameters for Atlantic halibut. Several critical classes of data pertaining to population and fishery dynamics are totally lacking for the Atlantic fishery (Table 5). As long as such shortages of information exist, the development of a more rational framework for management of Atlantic halibut will be difficult to attain. Priority areas for profitable future investigations include the determination of fishing mortality and stock identification.

Adequate information exists for effective management of the Pacific halibut fishery, and data sets for catch, effort, and age composition are among the most complete in the world. However, stock assessment and biological understanding would benefit from improved knowledge of quantitative trophic relationships,

Table 5. Status of data required for adequate management of halibut resources, western Atlantic and eastern Pacific

Type of study	Component	Atlantic	Pacific
Baseline biological	1. Trophic interactions	In progress	Qualitative studies done
	2. Early life history	None	Largely complete, with some work on juvenile migrations required
Stock identification	1. Tagging studies	Fairly complete	Fairly complete
	2. Morphometrics	None	Some, not published
	3. Genetic investigations	None	Some; otolith and DNA work started
	4. Spawning location	Study underway	Complete
Population parameters	1. Natural mortality (*M*)	Assumed < 0.2	Assumed = 0.2
	2. Fishing mortality (*F*)	Not known	Estimated from models
	3. Total mortality	Not known	Estimated from models
	4. Gear selection	Assumed equal to Pacific	Estimated from models, empirical study
	5. Age and growth	Sporadic, ≈ 1000/year	Extensive, > 15 000/year
	6. Recruitment	Not known	Estimated from models
	7. Population biomass	Not known	Estimated from models
Fishery statistics	1. Total landings	Reliable since 1963	Reliable since 1929
	2. Stock origin	Managed as separate stocks	Managed as single stock
	3. Age component by gear	Few data	Extensive data
	4. Effort	Limited information	Good data for major removals
	5. Discarding	Unknown, probably small	Observer estimates, included in models
	6. Survival of discards	Empirically estimated	Empirically estimated
	7. Changes in efficiency	Have occurred, effects unknown	Have occurred, effects estimated

juvenile migration, stock structure investigations, and parameters used in stock assessment models.

On the Pacific coast, trophic studies completed to date have been largely qualitative (Table 5). The lack of quantitative data is of concern not only for hindering a better understanding of the ecosystem in general but also for preventing progress in understanding the dynamics of multispecies fisheries. Under the general topic of early life history investigations, the Pacific coast fishery does not suffer from severe deficiencies, with the exception of studies related to post-metamorphosed juveniles. Of particular interest would be work leading to a recruitment index for juveniles and a better understanding of their migration. The latter investigation is viewed as a particularly high priority, given concerns that juveniles intercepted in Alaskan waters have a significant impact on Canadian fisheries off British Columbia.

New research is underway to rectify some deficiencies in our understanding of stock structure of Pacific halibut, including investigations utilizing some of the newer genetic techniques for stock discrimination. One such study will apply DNA analysis (Graves and Dixon 1986) to tissue samples collected throughout the Pacific halibut range.

Although the Pacific halibut fishery is supported by a comparative wealth of information, the rather ubiquitous problem of the estimation of natural mortality rate remains. While this problem is common to most fisheries, it is of particular concern for comparatively long-lived species such as halibut. A recent sensitivity analysis (McCaughran and Deriso 1988) showed that abundance estimates were particularly sensitive to the assumed value of the natural mortality rate, whereas annual surplus production was less sensitive. A further area identified for future research includes an examination of changes in weight at age, which often show quite pronounced variation, even from year to year.

Market Competition

Little information is published about marketing either of the halibuts, and most information comes from interviews with knowledgeable industry sources.[5] Both North American halibuts are highly prized fish in the market place but seem to have separate niches. On the whole, Atlantic halibut commands a higher ex-vessel price than Pacific halibut. Pacific halibut is a relatively large volume product with predictable quantities available on a time schedule set by IPHC regulations. Most product is frozen, and plays an important role for many end users. Until the individual quota program in Canada, those users who demand fresh Pacific halibut had to schedule around three or four major 1-day openings. Fresh Pacific halibut from the Canadian fishery is now available from March through October, mostly in spring and

[5] Contacts included Richard Lord, Fulton Fish Market, New York, N.Y.; Steve LaHaie, Shaws' Crab House Restaurant, Chicago, Ill.; Charles Kartstein, Slade Gorton Company, Boston, Mass.; Mark Sandvik, Icicle Seafoods, Seattle, Wash.; Jon Rowley, Fish Works!, Seattle, Wash.

fall to avoid competition with salmon. Atlantic halibut landings are an order of magnitude less than for Pacific halibut, but fresh product is available throughout the year. Atlantic landings increase during the summer months relative to winter, as the fish are more available and the weather is better. The low volume of Atlantic halibut landings makes supply sporadic, such that the product cannot be a marketing cornerstone except very near the points of supply.

Atlantic halibut has almost no impact on marketing activities for Pacific halibut: the large quantity of the latter generates its own supply and demand. During periods of maximum availability following a major opening, gluts tend to hold down ex-vessel price, and ex-vessel price tends to increase from lower values in Bering Sea and Aleutian Islands ports to higher values nearer the distribution centres of Seattle, Wash., and Vancouver, B.C. Price for Canadian Pacific halibut has generally exceeded the U.S. price, owing in part to lower quantities and to a perception that the quality of Canadian-caught halibut is superior. The difference appears to have increased since the individual quota program in Canada. Market price of Pacific halibut increases with distance from the west coast distribution centres. At markets near the U.S.-Canada border on the east coast, fresh Pacific halibut may be more expensive than Atlantic halibut because of transportation costs.

The small quantity of landings generally prevents Atlantic halibut from becoming a keystone item in the market but is rather treated as a specialty item. Very little Atlantic halibut is frozen, as the fresh market can absorb the entire production and the small size (5-10 kg) is not as desirable as a frozen product as are larger halibut. Atlantic halibut is marketed primarily on the east coast of the U.S. and Canada as a relatively rare and therefore expensive commodity. Price, however, drops during the summer when large quantities of fresh and frozen Pacific halibut are available and the availability of Atlantic halibut increases. Ex-vessel prices of Atlantic halibut have tended to be higher than for Pacific halibut over the last few years (Table 6).

Marketing of Atlantic and Pacific halibut could change dramatically if management of Pacific halibut incorporates a program (e.g., through limited access) that changes large volume Alaskan or Canadian landings during three or four fishing periods to smaller volumes over a longer period. Various segments of the industry are debating the effects of such a change.

Those who support the current system point out that Pacific halibut is primarily a frozen product, and that the market could not absorb the entire production in a fresh form. Under the current system, fresh and frozen product is available in predictable quantities at known times, which aids planning for special promotions. Proponents of status quo management claim that substantially increased availability of fresh product would cause a white fish glut that would decrease the value of halibut and similar species. Timing of the Pacific halibut openings is set to minimize conflicts with fishing for other species, which also minimizes marketing competition for those species, especially Pacific salmon.

Table 6. Average annual ex-vessel price (U.S. $/kg) of Atlantic and Pacific halibut

Year	Atlantic	Pacific
1974	2.07	1.54
1975	2.24	1.96
1976	1.87	2.77
1977	1.74	2.88
1978	1.85	3.74
1979	2.24	4.69
1980	2.11	2.18
1981	2.20	2.24
1982	2.24	2.40
1983	2.49	2.49
1984	2.46	1.65
1985	2.60	1.96
1986	3.52	3.17
1987	4.03	3.48
1988	4.62	2.82
1989	3.87	3.30
1990	9.15	4.14

Support for changed management comes from those who desire access to fresh halibut on regular basis, perhaps year-round, such as seafood retailers and restauranteurs. These users profess pent up demand for fresh halibut that would use a much larger share of the Pacific halibut resource than is currently available. Proponents for change often claim that quality of fresh and frozen Pacific halibut may suffer from poor handling during the short, intense fisheries. While both sides agree that overall quality is generally high, the incidence of poor quality could be reduced with less intense fishing.

Aquaculture

Although at first Atlantic halibut appeared to be a most unlikely candidate for aquaculture, five countries conduct active research on its development for aquaculture (Waiwood 1988). Norway took the lead in this research in the early 1980's. Others are Scotland, Iceland, Faroe Islands, and more recently Canada, with research programs in place at the Ocean Science Center, Memorial University, Nfld., and the DFO Biological Station, St. Andrews, N.B. Haug (1990) reviewed the rearing experiments, particularly in Norway, that are the foundation for aquaculture potential.

There are distinct advantages to cultured Atlantic halibut, such as rapid growth rate which continues into larger sizes, hardiness in surviving capture by commercial gears, adaptiveness to captivity, and high potential price. However, Waiwood (1988) also points to areas of considerable concern. These include reducing the high mortality in producing juveniles from the egg stage, reducing the possibility of infection and disease at young stages, providing an adequate and appropriate food supply, and providing the optimum combinations of physical and external conditions. Although investigators believe that Atlantic halibut remain a promising product of aquaculture, Waiwood (1988) reports that it will probably be at least 6 years before large-scale production in Europe could be expected. Norwegian aquaculture companies are currently marketing cultured halibut on a limited scale as production capacity expands (R. Engelson, SeaFarms A/S, Bergen, Norway, personal communication). In Atlantic Canada, Waiwood (1988) speculates that production capability could be realized in 8-10 years only if Canada's research effort and support to develop a comprehensive research program is increased significantly.

Research on spawning and raising Pacific halibut in captivity is underway in the U.S. at the University of Washington in Seattle and in Canada at the DFO laboratory in Nanaimo, B.C. These programs are designed to learn more about the early life history of the species, and no effort is directed toward aquaculture at this time. The emphasis is on obtaining and holding brood stock, inducing spawning and obtaining fertilized eggs, hatching the eggs and raising larvae. Fertilized eggs have been successfully obtained but with limited hatching success through 1988. Hatching success and survival greatly improved in 1989, but the fish did not spawn successfully in 1990.

Summary

North American halibut (*Hippoglossus* spp.) are important commercial species in both the Atlantic and Pacific Oceans. However, Atlantic halibut (*H. hippoglossus*) has been the subject of little research and management until recently, while Pacific halibut (*H. stenolepis*) has been managed and investigated by the International Pacific Halibut Commission since 1923. The conventional separation of the halibut into separate species is generally accepted, but it is a recommended topic for future research. The latitudinal range of the halibuts is the same in both oceans, but oceanographic dynamics cause the centre of the range to be more southerly in the Atlantic.

Spawning activity is the same for both halibuts, including spawning time and depth, egg size and time to hatching, and larval size. Atlantic halibut larvae, however, metamorphose and settle to the bottom at an earlier stage of development and smaller size than typical of Pacific halibut. Near surface currents carry Pacific halibut eggs and larvae in a consistent drift to the north and west along the coasts of British Columbia and Alaska, while no such uniformity is reported for Atlantic halibut larvae. Countermigration by Pacific halibut larvae to balance the effects of egg and larval drift is confirmed by tagging. Tagged juvenile Atlantic halibut have a more random distribution. The juvenile Pacific halibut countermigration is effectively completed near the time of sexual maturity, and Pacific halibut adults show little net migratory movements thereafter.

Adults of both halibuts migrate offshore to spawning grounds and inshore to return to feeding grounds. Males of both halibuts mature later and at a smaller size than do females, and males have a higher natural mortality. Pacific halibut and most Atlantic halibut exhibit sexual dimorphism, with females significantly larger than males of the same age. In contrast, Atlantic halibut from Newfoundland grow at the same rate for male and females. Both halibuts increase the proportion of fish in the diet with increasing age. Atlantic halibut smaller than 70 cm SL tend to have a diet with a higher proportion of invertebrates than do Pacific halibut, while at larger sizes, Atlantic halibut feed on a higher proportion of fish than do Pacific halibut.

Aboriginal fisheries on the Pacific coast gradually gave way during the 1800's to commercial fisheries using hook and longline gear set from dories carried by sailing vessels. A dory fishery in the Atlantic developed about the same time. The Pacific fishery continued hook and line gear to the present time with modern longline vessels, while the Atlantic fishery allowed trawl retention as well. Since the turn of the century, Atlantic landings have ranged from 1000 to 5000 t/year, while Pacific landings varied from 12 000 to 40 000 t/year. The IPHC has collected catch, effort, and age data by statistical and management areas along the coast since its inception in 1923. Catch data for the Atlantic halibut are most reliable since 1963 when Greenland halibut were recorded separately from Atlantic halibut. Other data, including catch per unit effort (CPUE), have been collected only intermittently for the Atlantic halibut. Intensive stock assessment analyses have been conducted for Pacific halibut, progressing from initial tracking of CPUE to catch-age analysis. Hypotheses for explaining Pacific halibut population control, invoking density dependence and exogenous effects, remain unresolved and as controversial as during the Thompson-Burkenroad debates of the 1930's and 1940's.

While Pacific halibut has been intensely managed since 1923, active management of the Atlantic halibut began in 1988. Provisional Atlantic quotas and a minimum size limit now prevail. Bycatches of halibut, the incidental harvest by fisheries targeting on other species, have much different implications in the Atlantic andPacific fisheries. While trawl retention is allowed in the Atlantic, Pacific halibut bycatch must be returned to the sea and bycatch mortality limits have been established that limit groundfish fisheries. The Atlantic fishery operates under limited entry with harvest allocations assigned by vessel size, but management is under pressure to increase entrants and the amount of quota. The Pacific fishery in the U.S., under open access, has seen substantial growth in the fleet and its catching capacity that causes difficulty in managing to the quotas, enforcement problems, and short seasons leading to gluts, quality problems, and potentially unsafe fishing operations. The Canadian Pacific halibut fishery has less problems because limited entry allowed a relatively smaller fleet, which operates under an individual quota system.

Acknowledgements

Forethought and insight of past and present directors and staff of the IPHC provided the historical background and data that formed the basis of this report. Without their contributions, we would have little basis for making comparisons. Several reviewers added organization and clarity for which we are grateful. S. H. Hoag and G. St-Pierre of the IPHC provided internal review. Dr. B. M. Leaman, Pacific Biological Station, Nanaimo, Dr. W. B. Scott, Huntsman Marine Science Center, St. Andrews, and B. E Skud, former director of IPHC, improved our efforts with constructive comments and challenges to incomplete documentation. P. Perley of the St. Andrews Biological Station assisted with data collection during research in the Atlantic. F. Cunninghan of the St. Andrews Biological Station and K. E. Exelby of the IPHC drew or revised many of the figures. The Maritime Museum of the Atlantic, Halifax, N.S., and the IPHC provided historical photographs. Ms. Hilary Stewart graciously allowed reproduction of her drawings of northwest Indian fishing methods. The Canadian Department of Fisheries and Oceans and the IPHC provided time and funds needed to complete this work, and the IPHC provided funds for publication.

References

Babcock, J. P., Found W. A., and O'Malley, H. 1931. Report of the International Fisheries Commission. Int. Fish. Comm. Rep. No. 1.

Bagenal, T. B. 1978. Aspects of fish fecundity. *In* Ecology of freshwater fish production. *Edited by* S. D. Gerking. John Wiley & Sons, New York, NY. pp. 75–101.

Beacham, T. D. 1983*a*. Variability in size and age at sexual maturity of witch flounder, *Glyptocephalus cynoglossus*, in the Canadian Maritimes region of the Northwest Atlantic Ocean. Can. Field-Nat. **97**: 409–422.

Beacham, T. D. 1983*b*. Variability in size and age at sexual maturity of American plaice and yellowtail flounder in the Canadian Maritimes region of the Northwest Atlantic Ocean. Can. Tech. Rep. Fish. Aquat. Sci. **1196**: 1–79.

Bell, F. H. 1981. The Pacific halibut, the resource and the fishery. Alaska Northwest Publishing Co., Anchorage, Alaska.

Best, E. A. 1968. Studies of young halibut: census of juveniles. West. Fish. **75**(5): 38–41, 59–60.

Best, E. A., and St-Pierre, G. 1986. Pacific halibut as predator and prey. Int. Pac. Halibut Comm. Tech. Rep. No. 21.

Bigelow, H. B., and Schroeder , W. C. 1953. Fishes of the Gulf of Maine. U.S. Fish Wildl. Serv. Fish. Bull. **53**: 1–577.

Boulva, S., and McLaren, I. A., 1979. Biology of the harbour seal, *Phoca vetulina*, in eastern Canada. Bull. Fish. Res. Board Can. **200**: 1–24.

Bowering, W. R. 1986. The distribution, age and growth and sexual maturity of Atlantic halibut (*Hippoglossus hippoglossus*) in the Newfoundland and Labrador area of the Northwest Atlantic. Can. Tech. Rep. Fish. Aquat. Sci. **1432**: 1–34.

Bowering, W.R. 1988. An analysis of morphometric characters of Greenland halibut (*Reinhardtius hippoglossoides*) in the Northwest Atlantic using a multivariate analysis of covariance. Can. J. Fish. Aquat. Sci. **45**: 580–585.

Brodeur, R. D., and Livingston, P. A. 1988. Food habits and diet overlap of various eastern Bering Sea fishes. NOAA Tech. Mem. NMFS F/NWC-127.

Cliche, G. 1986. Résultats de la pêche d'exploration au flétan atlantique (*Hippoglossus hippoglossus*) réalisée â Ile d'Anticosti en 1985. MAPAQQ-Pecheries DRST - Doc. rech. **86/8**: 1–29.

Cox, P. 1924. Larvae of the halibut (*Hippoglossus hippoglossus* L.) on the Atlantic coast of Nova Scotia. Contrib. Can. Biol. Fish. **1**(21): 411–412.

Deriso, R. B. 1980. Harvesting strategies and parameter estimation for an age-structured model. Can. J. Fish. Aquat. Sci. **37**: 268–282.

Deriso, R. B., and Quinn, T. J., II. 1983. The Pacific halibut resource and fishery in regulatory area 2. II. Estimates of biomass, surplus production, and reproductive value. Int. Pac. Halibut Comm. Sci. Rep. No. 67. pp. 55–89.

Deriso, R. B., Quinn, T. J. II, and Neal, P. R. 1985. Catch-age analysis with auxiliary information. Can. J. Fish. Aquat. Sci. **42**: 815–824.

Deriso, R. B., Hoag, S. H., and McCaughran, D. A. 1986. Two hypotheses about factors controlling production of Pacific halibut. Int. North Pac. Fish. Comm. Bull. 47. pp. 167–174.

Devold, F. 1938. The North Atlantic halibut and net fishing. Fiskeri Dir. Skr. Ser. Havundersr **5**: 1–47.

Drinkwater, K. 1987. "Sutcliffe revisited": previously published correlations between fish stocks and environmental indices and their recent performance. Can. Tech. Rep. Fish. Aquat. Sci. **1556**: 41–56.

Drinkwater, K. F., and Trites, R. W. 1987. Monthly means of temperature and salinity in the Scotian Shelf region. Can. Tech. Rep. Fish. Aquat. Sci. **1539**: iv + 1–101.

Fairbairn, D. J. 1981. Biochemical genetic analysis of population differentiation in Greenland halibut (*Reinhardtius hippoglossoides*) from the Northwest Atlantic, Gulf of St. Lawrence, and Bering Sea. Can. J. Fish. Aquat. Sci. **38**: 669–677.

Fevolden, S.E., and Haug, T. 1988. Genetic population structure of Atlantic halibut, *Hippoglossus hippoglossus* (L.). Can. J. Fish. Aquat. Sci. **45**: 2–7.

Forrester, C. R., and Alderdice, D. F. 1973. Laboratory observations on early development of the Pacific halibut. Int. Pac. Halibut Comm. Tech. Rep. No. 9. pp. 1–15.

Fredin, R. A. 1987. History of regulation of Alaskan groundfish regulations. U.S. Nat. Marine Fish. Serv. NWAFC Processed Rep. 87–07.

Freeberg, M. Regulations affecting bycatch in the North Pacific, 1923–1991. Natural Resource Consultants, Seattle, WA. In press.

Frost, K. J., and Lowry, L. F. 1981. Trophic importance of some gadids in northern Alaska and their body-otolith size relationships. Fish. Bull. **79**: 187–192.

Godø, O. R., and Haug, T. 1988. Tagging and recapture of Atlantic halibut, *Hippoglossus hippoglossus,* in Norwegian waters. J. Cons. Int. Explor. Mer, **44**: 169–179.

Goode, G. B., and Collins, J. W. 1887. The fresh-halibut fishery. *In* The fisheries and fishery industry of the United States. Section V. History and methods of the fisheries, Vol. I, Part I. The halibut fisheries. Government Printing Office, Washington, DC. pp. 3–89.

Grant, W. S., Teel, D. J., Kobayashi, T., and Schmitt, C. 1984. Biochemical population genetics of Pacific halibut (*Hippoglossus stenolepis*) and comparison with Atlantic halibut (*Hippoglossus hippoglossus*). Can. J. Fish. Aquat. Sci. **41**: 1083–1088.

Graves, J. E., and Dixon, A. E. 1986. Mitochondrial DNA genetic similarity of Atlantic and Pacific skipjack tuna and its management implications. *In* Proceedings of the ICCAT Conference on the International Skipjack Year Program. *Edited by* P. E. K. Symons, P. M. Miyake, and G. T. Sakagawa. International Commission for the Conservation of Atlantic Tunas, Madrid. pp. 237–241.

Grier, H. J. 1981. Cellular organization of the testis and spermatogenesis in fish. Am. Zool. **21**: 345–357.

Harkonen, T. 1986. Guide to the otoliths of the bony fishes of the northeast Atlantic. Danbiu ApS., Hellerup, Denmark.

Haug, T. 1990. Biology of the Atlantic halibut, *Hippoglossus hippoglossus* (L. 1758). Adv. Mar. Biol. **26**: 1–70.

Haug, T., and Sundby, S. 1987. A preliminary report on the natural occurrence and ecology of Atlantic halibut, *Hippoglossus hippoglossus*, postlarvae and young immature stages. ICES CM Doc. F:38.

Haug, T., Kjørsvik, E., and Solemdal, P. 1984. Vertical distribution of Atlantic halibut (*Hippoglossus hippoglossus*) eggs. Can. J. Fish. Aquat. Sci. **41**: 798–804.

Haug, T., Kjørsvik, E., and Solemdal, P. 1986. Influence of some physical and biological factors on the density and vertical distribution of Atlantic halibut *Hippoglossus hippoglossus* eggs. Mar. Ecol. Prog. Ser. **33**: 207–216.

Hoag, S. H. 1975. Survival of halibut released after capture by trawls. Int. Pac. Halibut Comm. Sci. Rep. No. 57.

Hoag, S. H., and McNaughton, R. J. 1978. Abundance and fishing mortality of Pacific halibut, cohort analysis, 1935–1976. Int. Pac. Halibut Comm. Sci. Rep. No. 65.

Homans, R. E. S., and Needler, A. W. H. 1944. The food of haddock. Proc. N.S. Inst. Sci. **21**: parts 1 and 2.

IPHC. 1985. Annual report, 1984. International Pacific Halibut Commission.

IPHC. 1987. The Pacific halibut: biology, fishery, and management. Int. Pac. Halibut Comm. Tech. Rep. No. 22.

IPHC. 1988. Annual report, 1987. International Pacific Halibut Commission.

IPHC. 1989. Annual report, 1988. International Pacific Halibut Commission.

Jákupsstovu, S. H., and Haug, T. 1988. Growth, sexual maturation, and spawning season of Atlantic halibut, *Hippoglossus hippoglossus*, in Faroese waters. Fish. Res. **6**: 201–215.

Jensen, A. G., and Wise, J. P. 1961. Movement of tagged halibut off New England II. Trans. Am. Fish. Soc. **90**: 489–490.

Jesperson, P. 1926. On the halibut in Icelandic waters. Rapp. P.-v. Reun. Cons. Int. Explor. Mer, **39**: 103–113.

Joensen, J. S. 1954. On the life history of halibut in Faroese waters. Medd. Komm. Fiskeri-og Hav. Ser. Fisk. **1** NS(5): 1–25.

Kask, J. L. 1935. Studies in migration fishing mortality and growth in length of the Pacific halibut (*Hippoglossus hippoglossus*) from marking experiments. Doctoral dissertation, University of Washington, Seattle, WA.

Khan, R. A., Dawe, M., Bowering, R., and Misra, R. K. 1982. Blood protozoa as an aid for separating stocks of Greenland halibut, *Reinhardtius hippoglossoides*, in the northwestern Atlantic. Can. J. Fish. Aquat. Sci. **39**: 1317–1322.

Kjørsvik, E., Haug, T., and Tjemsland, J. 1987. Spawning season of the Atlantic halibut (*Hippoglossus hippoglossus*) in northern Norway. J. Cons Int. Explor. Mer, **43**: 285–293.

Kohler, A. C. 1964. Movements of halibut on the Nova Scotian and Grand Banks. J. Fish. Res. Board Can. **21**: 837–840.

Kohler, A.C. 1967. Size at maturity, spawning season, and food of Atlantic halibut. J. Fish. Res. Board Can. **24**: 33–66.

Kolloen, L. N. 1934. Egg production in the Pacific halibut correlated with length, weight, and age. M.Sc. thesis, University of Washington, Seattle, WA.

Lear, W. H., and Misra, R. K. 1978. Clinal variation in scale characters of Atlantic salmon (*Salmo salar*) based on discriminant function analysis. J. Fish. Res. Board Can. **35**: 43–47.

Lønning, S., Kjørsvik, E., Haug, T., and Gulliksen, B. 1982. The early development of the halibut, *Hippoglossus hippoglossus*, (L.), compared with other marine teleosts. Sarsia, **67**: 85–91.

Martin, W. R., and McCracken, F. D. 1950. Movements of halibut on the Canadian coast. Fish. Res. Board Can. Prog. Rep. Atl. Coast Stn. No. 50. pp. 3–8.

McCaughran, D. A. 1981. Estimating growth parameters of Pacific halibut from mark–recapture data. Can. J. Fish. Aquat. Sci. **38**: 394–398.

McCaughran, D.A. 1987. Growth in length of Pacific halibut. *In* Age and growth of fish. *Edited by* R. C. Summerfelt and G. E. Hall. The Iowa State University Press, Ames, IA. pp. 507–515.

McCaughran, D.A., and Deriso, R. B. 1988. Effect of biological assumptions on halibut production estimates. *In* Lecture notes on coastal and estuarine studies. 28. Fishery science and management. *Edited by* W. S. Wooster. Springer-Verlag, Berlin and New York. pp. 5–29.

McCracken, F. D. 1958. On the biology and fishery of the Canadian Atlantic halibut, *Hippoglossus hippoglossus* L. J. Fish. Res. Board Can. **15**: 1269–1311.

McCracken, F. D., and Martin, W. R. 1955. Recent recoveries of tagged halibut. Fish Res. Board Can. Prog. Rep. Atl. Coast Stn. No. 61. pp. 3–4.

McFarlane, G. A., Jensen, J. O. T., Andrews, W. T., and Groot, E. P. 1991. Egg and yolk sac development of Pacific halibut (*Hippoglossus stenolepis*). Int. Pac. Halibut Comm. Tech. Rep. No. 24.

McIntyre, A. D. 1953. The food of halibut from North Atlantic fishing grounds. Mar. Res. (Scotland), **1952**: 1–20.

McKenzie, R. A. 1946. The Canadian Atlantic halibut fishery. Bull. Fish. Res. Board Can. **71**: 1–29.

Misra, R. K., and Bowering, W. R. 1984. Stock delineation of Greenland halibut in the Northwest Atlantic using a recently developed, multivariate statistical analysis based on meristic characters. N. Am. J. Fish. Manage. **4**: 390–398.

Mulligan, T. J., Martin, F. D., Smucker, R. A., and Wright, D. A. 1987. A method of stock identification based on the elemental composition of striped bass *Morone saxatilis* (Walbaum) otoliths. J. Exp. Mar. Biol. Ecol. **114**: 241–248.

Myhre, R. J. 1974. Minimum size and optimum age of entry for Pacific halibut. Int. Pac. Halibut Comm. Sci. Rep. No. 55.

Myhre, R. J., and Quinn, T. J., II. 1984. Comparison of efficiency of snap gear to fixed-hook setline gear for catching Pacific halibut. Int. Pac. Halibut Comm. Sci. Rep. No. 69.

Myhre, R. J., Peltonen, G. J., St-Pierre, G., Skud, B. E., and Waldon, R. E. 1977. The Pacific halibut fishery: catch effort and CPUE, 1929–1975. Int. Pac. Halibut Comm. Tech. Rep. No. 14.

Neilson, J. D., and Bowering, W. R. 1989. Minimum size regulations and implications for yield and value in the Canadian Atlantic halibut (*Hippoglossus hippoglossus*) fishery. Can. J. Fish. Aquat. Sci. **46**: 1899–1903.

Neilson, J. D., Bowering, W. R., and Frechet, A. 1987. Management concerns for Atlantic halibut (*Hippoglossus hippoglossus*) in the Canadian North Atlantic. CAFSAC Res. Doc. 87/73.

Neilson, J. D., Waiwood, K. G., and Smith, S. J. 1989. Survival of Atlantic halibut (*Hippoglossus hippoglossus*) caught in longline and otter trawl gear. Can. J. Fish Aquat. Sci. **46**: 887–897.

Neilson, J. D., Kearney, J. F., Perley, P., and Sampson, H. 1993. Reproductive biology of Atlantic halibut (*Hippoglossus hippoglossus*) in Canadian waters. Can. J. Fish Aquat. Sci. In press.

Norman, J.R. 1934. A systematic monograph of the flatfishes (Heterosomata). Vol. 1. The British Museum, London.

Novikov, N. P. 1964. Basic elements of the biology of the Pacific halibut (*Hippoglossus hippoglossus stenolepis* Schmidt) in the Bering Sea. *In* Soviet fisheries investigations in the Northeast Pacific. Part II. *Edited by* P.A. Moiseev. Israel Program for Scientific Investigations. pp. 175–219.

Øiestad, V., and Haugen, A. S. 1980. Rearing of halibut larvae (*Hippoglossus hippoglossus* L.) to metamorphosis and beyond. ICES CM 1980/F:9.

Parma, A. M. 1990. Optimal harvesting of fish populations with non-stationary stock-recruitment relationships. Nat. Resour. Model. **4**(1): 39–76.

Parma, A. M., and Deriso, R. B. 1990. Experimental harvesting of cyclic stocks in the face of alternative recruitment hypotheses. Can. J. Fish. Aquat. Sci. **47**(3): 595–610.

Peeling, D. A., and Rodgers, D. M. 1984. Circle hook comparison study. Proj. Rep. No. 78, Fish. Dev. Br., Dep. Fish. Oceans, Halifax, NS.

Pittman, K., Berg, L., and Naas, K. 1987. Morphological development of halibut (*Hippoglossus hippoglossus*) larvae with special reference to mouth development and metamorphosis. ICES CM 1987/F:18.

Pope, J. G. 1972. An investigation of the accuracy of virtual population analysis using cohort analysis. Int. Comm. Northwest Atl. Fish. Res. Bull. No. 9. pp. 65–74.

Quinn, T. J. II., Deriso, R. B., and Hoag, S. H. 1985. Methods of population assessment of Pacific halibut. Int. Pac. Halibut Comm. Sci. Rep. No. 72.

Rae, B. B. 1958. The occurrence of plerocercoid larvae of *Grillotia erinaoeus* (vanBeneden) in halibut. Mar. Res. (Scotland), **1958**: 1–31.

Schmitt, C. C., and Skud, B. E. 1978. Relation of fecundity to long-term changes in growth, abundance and recruitment. Int. Pac. Halibut Comm. Sci. Rep. No. 66.

Scott, J. S., and Bray, S. A. 1989. Helminth parasites of the alimentary tract of Atlantic halibut (*Hippoglossus hippoglossus* L.) and Greenland halibut (*Reinhardtius hippoglossoides* (Walbaum)) on the Scotian shelf. Can. J. Zool. **67**: 1476–1481.

Scott, T. 1910. On the food of the halibut with notes on the food of *Scorpaena, Phycis blennoides*, the garpike and *Chimaera monstrosa*. 28th Report of the Fisheries Board, Scotland, 1909(3): 24–37.

Scudder, N. P. 1887. The salt-halibut fishery. *In* The fisheries and fishery industry of the United States. Section V. History and methods of the fisheries, Vol. I, Part I. The halibut fisheries. Government Printing Office, Washington, DC. pp. 90–119.

Shaklee, J. B., Tamaras, C. S., and Waples, R. S. 1982. Speciation and evolution of marine fishes studied by electrophoretic analysis of proteins. Pac. Sci. **36**: 141–157.

Sigurdsson, A. 1956. Contribution to the life history of the halibut at the west of Iceland in recent years (1936–1950). Medd. Komm. Fisk. Havunders. Ser. Fisk. **1** NS(5): 1–24.

Skud, B. E. 1975. Revised estimates of halibut abundance and the Thompson-Burkenroad debate. Int. Pac. Halibut Comm. Sci. Rep. No. 56.

Skud, B. E. 1977. Drift, migration, and intermingling of Pacific halibut stocks. Int. Pac. Halibut Comm. Sci. Rep. No. 63.

Southward, G. M. 1962. A method of calculating body lengths from otolith measurements for Pacific halibut and its application to Portlock-Albatross grounds data between 1935 and 1957. J. Fish. Res. Board Can. **19**: 339–362.

Southward, G. M. 1967. Growth of Pacific halibut. Int. Pac. Halibut Comm. Rep. No. 43.

Stewart, H. 1977. Indian Fishing. Early methods on the Northwest Coast. University of Washington Press, Seattle, WA.

Stobo, W. T., Neilson, J. D, and Simpson, P.G. 1988. Movements of Atlantic halibut (*Hippoglossus hippoglossus*) in the Canadian North Atlantic. Can. J. Fish. Aquat. Sci. **45**: 484–491.

St-Pierre, G. 1984. Spawning locations and season for Pacific halibut. Int. Pac. Halibut Comm. Sci. Rep. No. 70.

St-Pierre, G. 1989. Recent studies of Pacific halibut postlarvae in the Gulf of Alaska and eastern Bering Sea. Int. Pac. Halibut Comm. Sci. Rep. No. 73.

St-Pierre, G. 1992. Visual determination of sex in live Pacific halibut. ICES J. Mar. Sci. **49**: 373–376.

Sullivan, P.J., Trumble, R.J., Alderstein, S.A. Pacific halibut bycatch in the groundfish fisheries: effects and management implications to the halibut fishery, Int. Pac. Halibut Comm. Sci. Rep., in press.

Sutcliffe, W. H., Jr. 1973. Correlations between seasonal river discharge and local landings of American lobster (*Homarus americanus*) and Atlantic halibut (*Hippoglossus hippoglossus*) in the Gulf of St. Lawrence. J. Fish. Res. Board Can. **30**: 856–859.

Tåning, A. V. 1936. On the eggs and young stages of the halibut. Medd. Komm. Fiskeri-og Havunders, Ser. Fisk. **10**(4): 1–23.

Tåning, A. V. 1938. Migration of small halibut marked in Faroese waters. J. Cons. Int. Explor. Mer, **13**: 370–375.

Thompson, W. F. 1915. A preliminary report on the life history of the halibut. Rep. of the BC Com. Fish., 1914. pp. N76–N99.

Thompson, W. F. 1916. The problem of the halibut. Rep. of the BC Com. Fish., 1915. pp. S130–S140.

Thompson, W. F., and Bell F. H. 1934. Biological statistics of the Pacific halibut fishery. 2. Effect of changes in intensity upon total yield and yield per unit of gear. Int. Fish. Comm. Rep. No. 8.

Thompson, W. F., and Herrington, W. C. 1930. Life history of the Pacific halibut. 1. Marking experiments. Int. Fish. Comm. Rep. No. 2.

Thompson, W. F., and Van Cleve, R. 1936. Life history of the Pacific halibut. 2. Distribution and early life history. Int. Fish. Comm. Rep. No. 9.

Trippel, E. A., and Harvey, H. H. 1991. A comparison of methods used to estimate age and length of fishes at sexual maturity using populations of white sucker (*Catostomus commersoni*). Can. J. Fish. Aquat. Sci. **48**: 1446–1459.

Trumble, R. J. 1992. Looking beyond time-area management of bycatch—an example from Pacific halibut. *In* Proceedings of the National Industry Bycatch Workshop. *Edited by* R. W. Schoning, R. W. Jacobson, D. L. Alverson, T. H. Gentle, and J. Auyong. Natural Resource Consultants, Seattle, WA. pp. 142–158.

Trumble, R. J., McGregor, I.R., St-Pierre, G., McCaughran, D. A., and Hoag, S. H. 1990. Sixty years of tagging Pacific halibut: a case study. Proceedings of the International Symposium and Educational Workshop on Fish Marking Techniques. Am. Fish. Soc. Symp. **7**: 831–840.

Trumble, R. J., St-Pierre, G., and. McGregor, I. R 1991. Evaluation of Pacific halibut management for Regulatory Area 2A. I. Review of the Pacific halibut fishery in Area 2A. Int. Pac. Halibut Comm. Sci. Rep. No. **74**: 7–39.

Tsuyuki, H., Roberts, E. and Best, E. A. 1969. Serum transferrin systems and the hemoglobins of the Pacific halibut (*Hippoglossus stenolepis*). J. Fish. Res. Board Can. **26**: 2351–2362.

Van Cleve, R., and Seymour, A. H. 1953. The production of halibut eggs on the Cape St. James spawning bank off the coast of British Columbia 1935-1946. Int. Fish. Comm. Rep. No. 19.

Vernidub, M.F. 1936. Materials contributing to the knowledge of Pacific halibut. Leningrad Soc. Nat. **65**(2): 143–182. (Int. Pac. Halibut Comm. transl.)

von Brandt, A. 1972. Fish catching methods of the world. Fishing News, London.

Waiwood, K. G. 1988. Halibut: a potential aquaculture species for Atlantic Canada. Aquat. Assoc. Can. Bull. **89**(2): 21–24.

Wallace, R. A., and Selman, K. 1981. Cellular and dynamic aspects of oocyte growth in teleosts. Am. Zool. **21**: 325–343.

Williams, G. H., Schmitt, C. C., Hoag, S. H., and Berger J. D. 1989. Incidental catch and mortality of Pacific halibut, 1962–1986. Int. Pac. Halibut Comm. Tech. Rep. No. 23.

Wise, J. P., and Jensen, A. G. 1959. Movement of tagged halibut off New England. I. Trans. Am. Fish. Soc. **88**: 357–358.

Zwanenburg, K. 1984. The Atlantic halibut. Underwater world. Department of Fisheries and Oceans, Canada.